# Science Education Standards

Edited by Eleanor D. Siebert and William J. McIntosh

D1308349

NATIONAL SCIENCE TEACHERS ASSOCIATION

**Art and Design**
Linda Olliver, Director
**NSTA Web**
Tim Weber, Webmaster
**Periodicals Publishing**
Shelley Carey, Director
**Printing and Production**
Catherine Lorrain-Hale, Director
**Publications Operations**
Erin Miller, Manager
*sci*LINKS
Tyson Brown, Manager

**National Science Teachers Association**
Gerald F. Wheeler, Executive Director
David Beacom, Publisher

NSTA Press, NSTA Journals,
and the NSTA Web site deliver
high-quality resources for
science educators.

Shirley Watt Ireton, Director
Beth Daniels, Managing Editor
Judy Cusick, Associate Editor
Jessica Green, Assistant Editor
Linda Olliver, Cover Design

*College Pathways to the Science Education Standards*
NSTA Stock Number: PB161X
ISBN 0-87355-193-1
Library of Congress Control Number: 2001088202
Printed in the USA by Kirby Lithographic Company, Inc.
Printed on recycled paper

# Table of
# Contents

**About the Editors** ................................................................... vii

**Preface** ................................................................................. viii

**Acknowledgments** ................................................................. xi

**Introduction** A Vision of Scientific Literacy .......................... xii

## Chapter 1

**Teaching Standards** *William J. McIntosh* ............................. 1

> **Standard A:** Planning an Inquiry-Based Science Program ................. 2
>
> From the Field: Changing Student Attitudes about Science
> through Inquiry-Based Learning—*Gerald H. Krockover* ..................... 3
>
> **Standard B:** Guiding and Facilitating Learning ............................... 4
>
> From the Field: Identifying and Using Learning Styles to Facilitate
> Instruction—*Nannette Smith* ................................................. 6
>
> **Standard C:** Linking Assessing, Learning, and Teaching .................... 8
>
> From the Field: Assessment Techniques to Guide Teaching in
> Courses with Large Enrollments—*Harry L. Shipman* ..................... 9
>
> **Standard D:** Designing and Managing the Learning Environment ... 12
>
> From the Field: Developing Experimental Design Skills of
> Students—*Patrick Gleeson* ................................................ 13
>
> **Standard E:** Building Learning Communities .............................. 15
>
> From the Field: Creating a Learning Community in Introductory
> Biology—*Marvin Druger* .................................................... 16
>
> **Standard F:** Participating in Program Development ....................... 18
>
> From the Field: Aligning Courses for Standards-Based Teaching—
> *Dana L. Zeidler* ............................................................ 20

# Chapter 2

## Professional Development Standards *Joseph I. Stepans, Maureen Shiflett, Robert E. Yager, and Barbara Woodworth Saigo* ................. 25

**Standard A:** Learning Science Content ........................................... 27

From the Field: Two Perspectives of a Workshop Experience—
*Maureen Shiflett* ...................................................................... 29

**Standard B:** Learning How to Teach Science ................................. 30

From the Field: A Science Program for Prospective Elementary
Teachers—*Joseph I. Stepans* ...................................................... 31

**Standard C:** Learning to Learn ...................................................... 35

From the Field: Research-Based Change: How One College
Professor Approached the Challenge of Changing Teaching—
*Diane Ebert-May* ...................................................................... 36

From the Field: Analysis of an Innovative College Chemistry
Course—*Barbara Woodworth Saigo* and *Susan Millar* ................. 39

**Standard D:** Planning Professional Development Programs ............. 41

From the Field: The Wyoming TRIAD (WyTRIAD) Professional
Development Process—*Barbara Woodworth Saigo* and *Joseph I. Stepans* .............. 47

From the Field: Preparing Future Faculty—*David Seybert* ............... 51

# Chapter 3

## Assessment Standards *Judith E. Heady, Brian P. Coppola, and Lynda C. Titterington* ................................................................. 57

**Standard A:** Coordination of Assessment with Intended Purposes .. 60

From the Field: Structured Study Groups (SSGs): Using Peer-Facilitated
Instruction to Develop Self-Assessment Skills—*Brian P. Coppola* ...................... 63

From the Field: The Counterintuitive Event: A Performance-Based Assessment—
*Brian P. Coppola* ...................................................................... 65

**Standard B:** Measuring Student Achievement and
Opportunity to Learn ................................................................. 68

From the Field: Using Journals to Assess Student Understanding
of Anatomy and Physiology—*Lynda C. Titterington* ..................... 70

From the Field: Teaching without Exams through the Use of
Student-Generated Portfolios in an Undergraduate Environmental
Geology Class—*Kent S. Murray* ................................................ 72

**Standard C:** Matching Quality of Data to Consequences ............... 74

From the Field: Effective Use of Pretests and Posttests—
*Judith E. Heady* ...................................................................... 76

From the Field: Using Assessment in Curriculum Reform—
*Gail Schiffer, Ben Golden, Gary Lewis,* and *Diane Willey* .................................. 78

**Standard D:** Avoiding Bias .............................................................. 80

From the Field: Using Student Strengths to Develop Assessment
Tools for Nonscience Majors—*Suzanne Shaw Drummer* ............................... 82

From the Field: Literature-Based Examinations and Grading Them:
Well Worth the Effort—*Brian P. Coppola* .................................................... 84

**Standard E:** Making Sound Inferences ........................................... 86

From the Field: Assessment as Student Learning—*Leona Truchan, George Gurria,*
and *Lauralee Guilbault* ...................................................................... 87

From the Field: Assessment Criteria as a Heuristic for Developing Student
Competency in Analysis and Evaluation of Published Papers in the Sciences—
*Robert A. Paoletti* .................................................................................. 89

# Chapter 4

**Content Standards** *Eleanor D. Siebert* ................................................ 95

**Standard A:** Science as Inquiry ..................................................... 97

From the Field: A Research Approach to the General Chemistry
Laboratory—*Eleanor D. Siebert* ............................................................. 98

**Standards B, C, D:** Subject Matter Content ................................... 99

From the Field: Rethinking the Content in Nonmajors' Science
Courses—*Mario W. Caprio* .................................................................. 101

**Standard E:** Science and Technology ............................................ 104

From the Field: Issues in Science, Technology, and Society—
*Gordon Johnson* ................................................................................. 105

**Standard F:** Science in Personal and Social Perspectives ................ 107

From the Field: Value-Based Science—Chemistry and the
Environment—*Theodore D. Goldfarb* ..................................................... 108

**Standard G:** History and Nature of Science ............................. 110

From the Field: Project Inclusion: Using the History of Diverse Cultures
to Facilitate the Teaching of Chemistry—*Janan M. Hayes* and *Patricia L. Perez* .... 111

# Chapter 5

**Science Education Program Standards** *Eleanor D. Siebert* ....... 115

**Standard A:** Program Consistency ............................................. 116

From the Field: Program Reviews: A Vital Component for Quality
Undergraduate Science Education—*Eleanor D. Siebert* ............................... 117

**Standard B:** Curriculum Criteria ............................................... 120

From the Field: Science for All Students: Taking This Mandate
Seriously—*Robert W. Harrill* ................................................................. 120

**Standard C:** Science and Mathematics ....................................... 124

From the Field: Linking Mathematics and Science Learning: The Long Island
Consortium for Interconnected Learning—*Jack Winn* ............................... 125

**Standard D:** Resources to Support the Science Program ................ 127

From the Field: Merck Support of Undergraduate Science Education—
*Susan K. Painter* .......................................................................... 128

**Standard E:** Opportunity to Learn .............................................. 131

From the Field: Equity in Science Opportunities at the
Undergraduate Level—*Elizabeth T. Hays* ............................................ 132

**Standard F:** Communities of Support for Teachers ....................... 134

From the Field: Building Natural Science Faculty
Communities—*Jeanne L. Narum* ....................................................... 135

# Chapter 6

**Science Education System Standards** *Mario W. Caprio* ........... 139

**Standard A:** Sharing a Vision ..................................................... 146

From the Field: Strength in Numbers: Uniting the Fronts of Higher
Education (Summary of Symposium)—*Brian P. Coppola* ............................ 147

**Standard B:** Coordination of Science Education Policies .............. 150

From the Field: Cost-effective Biology for Elementary Education Majors—
*Douglas Schamel* and *Leslie Gordon* ................................................. 152

**Standard C:** Sustained Policies .................................................. 155

From the Field: Nurturing Meaningful Relationships in Science
Education—*Stacy Treco* ................................................................. 156

**Standard D:** Resources for Change .............................................. 158

From the Field: Resources for Change: Educational Activities of the American
Chemical Society—*Stanley H. Pine* .................................................... 159

**Standard E:** Equitable Policies ................................................... 161

**Standard F:** Policy Review for Unintended Effects ....................... 162

**Standard G:** Individual Responsibility ........................................ 163

From the Field: Postsecondary Teachers of Science: Catalysts for Change—
*Mario W. Caprio* .......................................................................... 164

**Epilogue:** Making Science Education Available to All
*William H. Leonard* ...................................................................... 169

**Profiles of Contributors** ............................................................ 173

**Index** ...................................................................................... 185

# About the Editors

**Eleanor D. Siebert** (Ph.D., University of California, Los Angeles) chairs the Department of Physical Sciences and Mathematics and is a professor of chemistry at Mount St. Mary's College in Los Angeles. She teaches introductory physical science, chemistry, and physics to both science majors and nonmajors and instrumental analysis and thermodynamics at the upper-division level. Her research activities with undergraduates involve studies of phase separation in model biological systems. She is author of *Experiments for General Chemistry* and *Foundations for General Chemistry*; she has co-edited *Methods of Effective Teaching and Class Management for University and College Teachers of Science*. She is past president of the Society for College Science Teachers and past college director of the National Science Teachers Association. Currently she chairs the Committee on Public Relations and Communications for the American Chemical Society. She is listed in *Who's Who Among America's Teachers* and in *American Men and Women in Science*. e-mail: *esiebert@msmc.la.edu*

**William J. McIntosh** (Ph.D., Temple University) is a professor of science education at Delaware State University, where he teaches courses in introductory physical and Earth sciences as well as science methods courses. He is interested in the role of college and university faculty in systemic reform; he has pioneered changes in his own science courses and works collaboratively with faculty from other institutions to promote course reform. He is past president of the Society for College Science Teachers and serves on the Board of Directors for the National Science Teachers Association as director of the college division. e-mail: *bmcintsh@udel.edu*

# Preface

**Wanted:** *College and university science teachers wishing to become engaged in a comprehensive, important, and potentially transforming educational movement. Those who accept the challenge will join with K–12 teachers in a quest to give every American an essential understanding of the physical and biological processes that characterize our world, and to nurture curiosity and scientific habits of mind. In the process, all participants will experience change and renewal.*

A job announcement such as the above might describe what is in store for higher education faculty who internalize the principles and practices recommended by the *National Science Education Standards*. Although the *Standards* were written to specifically address science education in grades K–12, the job of transforming science education and achieving a scientifically literate public is so important that the *Standards* have special significance for higher education science faculty as well. As postsecondary teachers of science, we cannot be complacent in the belief that the way we have been teaching will empower our students to function effectively in the world of today and tomorrow.

Attention is focused on the need for change in science education because so many Americans—from children to adults—do not have a correct understanding of basic science concepts and processes and are, therefore, ill-equipped to make critical decisions in areas such as health and medicine, the environment, and biotechnology. More sadly, many are unable to appreciate the intricate wonder that is our Earth. Another driving force of the reform movement is the accelerating need for individual and national competence and competitiveness in a global economy that is increasingly based on science and technology. Mediocre science scores on international achievement tests such as the Third International Mathematics and Science Study (TIMSS) underscore the need for a comprehensive approach to changing the way that science is taught and learned.

The *National Science Education Standards* provide a comprehensive guide to move us toward achieving a scientifically literate nation. These standards are based on our current, research-based understanding of the teaching-and-learning process, common perceptions (including misperceptions) of scientific concepts, and the role of prior knowledge in learning. Hundreds of teachers, scientists, science educators, and administrators from across the country collaborated on the *National Science Education Standards,* which suggest both what students should know and be able to do at each developmental level and how we can align curriculum, instruction, and assessment to help them achieve these expectations. The *Standards* also address systemic issues of teacher preparation, professional development, the quality of school programs, and the entire educational system as a context for K–12 reform.

University and college professors of science are an integral part of this educational system because it is, in very large part, from our courses that society will learn its science. The lessons and experiences we provide will be passed to future generations—by way of our majors who enter fields of science and technology and by way of those nonmajors who make policy and those who approve it. The *Standards* ask that we approach this task differently than we have in the past.

We must also note the important role we play regarding our students who are preparing to become K–12 teachers. The responsibility of preparing teachers lies primarily with higher education, and here faculty members in both science and education have significant roles. The responsibility of science faculty members is to develop not only the science knowledge of our students, but also their understanding of the nature of science, their ability to understand and use scientific ways of thinking, and their ability to make connections and apply what they know to the world outside the science classroom. The responsibility of education faculty members is not only to provide fundamental information and skills related to teaching and learning, but also to mentor teachers in their ability to actively study and reflect on what they do and use their own research to make informed decisions about the appropriateness of curriculum, instruction, and assessment in their own classrooms.

The purpose of this book is to present and interpret the *Standards* in ways that are meaningful to higher education faculty members, especially those who teach science. The Teaching Standards and the Assessment Standards (Chapters 1 and 3) are the focal points of the book. The Professional Development Standards (Chapter 2) as presented in this book carry a dual message. First, they speak to science faculty about ways to develop their own teaching skills so as to maximize learning opportunities for students; second, they serve as a guide to faculty members who are involved in providing professional development for others, emphasizing deep learning and genuine conceptual understanding rather than superficial exposures. The Content Standards (Chapter 4) are foundations upon which college instruction can build. The Science Education Program Standards (Chapter 5) articulate criteria that can be used to create excellent science programs at the postsecondary level, and the Science Education System Standards (Chapter 6) consider external factors that affect science program development and implementation.

In each chapter, short essays address the implications for college science teaching of each Standard, and there are detailed illustrations—"From the Field"—of how *Standards*-based teaching is being implemented in undergraduate science courses. For readers interested in pursuing further the concepts discussed in this book, we provide reference lists in each chapter.

Throughout the book, we have tried to maintain the original emphasis of the *National Science Education Standards* and to convey their central vision. We agree that at the college level as well as in K–12 *all* students can learn science; consequently, we include suggestions for teaching nonscience majors and students with special needs. We also believe that science should be an active process that engages students both intellectually and physically—*especially* at the introductory level. To illustrate this approach, we have asked colleagues to share ways they have structured courses for maximum student involvement.

*College Pathways to the Science Education Standards* is full of ideas, examples, and suggestions. As such, it is an appropriate resource for many audiences. For example, while the book is directed explicitly to postsecondary teachers of science and science educators, it will be useful to high school teachers of advanced placement courses and to preservice teachers and graduate students as they begin their classroom careers. There is no one way to use this resource. It should raise questions and stimulate thinking about what and how we teach, and how we might study and improve what we do. We hope it will elevate consideration of the implications of the *National Science Education Standards* for postsecondary science teaching and, in so doing, encourage readers to develop and grow in their understanding of the teaching-and-learning process.

# Acknowledgments

This National Science Teachers (NSTA) publication on the impact of the *National Science Education Standards* on higher education has been a long time coming; it was conceived by the 1996 NSTA Committee on College Science Teaching. Chairperson Gerald Krockover and Bill McIntosh, then president of the Society for College Science Teachers, agreed to an overall plan for the publication, and Bill became editor of the book. The 1997-99 Committee on College Science Teaching continued the project with its chairperson, Eleanor Siebert, becoming co-editor of the project. The 1999-2000 committee members (Bill McIntosh, chairperson) served as readers of the document as the manuscript developed.

This book has truly been a labor of love. Ultimately its value stems from the vignettes provided by the more than forty dedicated science teaching faculty and others who are profiled at the end of the book. Thanks to our chapter coordinators for their persistence in soliciting articles in their areas of expertise: Joseph Stepans, a member of the 1996 College Committee, who coordinated the contributions to the chapter on professional development; Judith Heady, who culled many references and gathered many vignettes pertaining to assessment; Mario Caprio, who brought together many resources and perspectives in his chapter on the higher education system; and Bill Leonard, who provided the Epilogue.

The names of many members of the NSTA College Committees who contributed to the book are included within these pages; others include Daniel Domin, George Randall, and Jeffery Schultz, whose ideas are included in the text and who served as readers during the development phase of the manuscript. We have benefited greatly from the careful review and suggestions of Donald French, Eric Packenham, Harold Pratt, and Jeffrey Weld.

We are deeply grateful for the initial support and suggestions of Phyllis Marcuccio, former associate executive director of NSTA, and *especially* for the support and guidance of Shirley Watt Ireton, director of NSTA Press. Thanks, too, for the patience and outstanding work of our editor, Judy Cusick, who shepherded us and the manuscript to completion.

Also at the NSTA, Linda Olliver designed the cover and the book's layout; Jack Parker, Nguyet Tran, and Catherine Lorrain-Hale handled production; and Catherine Lorrain-Hale coordinated the printing of the book. Beth Daniels, Anne Early, Jessica Green, and Claudia Link provided additional, invaluable assistance.

*Eleanor D. Siebert*                                          *William J. McIntosh*

## Introduction

# A Vision of Scientific Literacy

*Scientific literacy is the knowledge and understanding of scientific concepts and processes required for personal decision-making, participation in civic and cultural affairs, and economic productivity. (NRC 1996, 22)*

## The Nature of Science

Science provides a way for us to understand the world in which we live. Scientists ask questions, then search for answers to those questions. While there are many pathways that can be taken in the search for an answer to a particular question, for scientists all ways are based on observations, inferences, and testing. Referred to as "the scientific method," these processes are the hallmarks of a method that produces creative and valid interpretations of events. During questioning for jury service recently, a judge asked one of the editors of this book whether she would be able to contend with uncertainties since she had been trained as a scientist to deal objectively with evidence about which there is no uncertainty. This judge's impression was that scientists work with the "black and white" issues of our world, while human judgments in dispensing justice must enter a "gray" area.

This story confirms that members of the public, including those who are well educated, often do not understand the nature of science. It has been well documented that many people do not comprehend fundamental concepts and processes of science and thus cannot appreciate the work that scientists do. Why is an understanding and appreciation of science important? In 1983, a report prepared by the National Commission on Excellence in Education (NCEE) entitled *A Nation at Risk: The Imperative for Educational Reform* (NCEE 1983) focused on the importance of science in developing and sustaining economic growth in the United States. Perhaps even more importantly, John A. Moore has argued that without an understanding of science, it is virtually impossible to appreciate the beauty and wonder of the world around us (Moore 1993). Finally, in a society where life is increasingly dominated by technology enabled by scientific discoveries, it is important to have a scientifically literate society—a populace that can make intelligent personal and political decisions.

If an understanding and appreciation of science is important to quality of life and to sustaining economic growth, how do we ensure a scientifically literate populace? Many argue that a cultural shift is required. One place to begin that shift is in school classrooms across the nation. Students must come to understand the nature of science, be familiar with the fundamental concepts that connect the science disciplines, and know how to apply those concepts in the process of making decisions. This cultural shift will take years to attain and will happen only if there is a public consensus that the goal is valid—and only if the educational community has a shared vision for how to reach that goal.

## Changing the Way That Science Is Taught in the K–12 Classroom

In *A Nation at Risk* (NCEE 1983), the National Commission on Excellence in Education predicted that the United States would soon be engulfed in a "rising tide of mediocrity." The report provided a rationale and background for program shifts in funding priorities at the National Science Foundation. Education programs that emerged included the Eisenhower Program and systemic reforms. The report was followed in 1989 by President George Bush's national education meeting at the National Governors' Conference. The governors set national education goals for science, technology, and mathematics; Goal 4 stated that by the year 2000, children in the United States would rank number one in understanding science and mathematics. Unfortunately, a coherent national plan to achieve these goals was not developed at that time.

In 1989, the American Association for the Advancement of Science (AAAS) published *Science for All Americans* (AAAS 1989), a study that argued that *all* students could be and should be expected to learn science. This initial publication of AAAS's Project 2061 marked the adoption of a long-range plan for achieving scientific literacy. In 1993, AAAS published *Benchmarks for Science Literacy*, which outlined the hallmarks of a plan to achieve scientific literacy in the schools. In 1989 the National Science Teachers Association (NSTA) began work on a curriculum project called

"Scope, Sequence, and Coordination of Secondary School Science," which integrated concepts across science disciplines to build a coherent understanding of science for children in the middle grades; the project was tested extensively at that level across the nation. Its success is documented by the National Science Teachers Association (NSTA 1992) and Aldridge (1996), although it did not achieve nationwide acceptance.

Most of these projects are still in progress, but a more comprehensive approach was needed in order to provide a coherent national vision for science education, particularly at the K–12 level. In 1992 the National Research Council of the National Academy of Sciences convened a committee to consider the articulation of national standards for science education. Three working groups, composed of teachers, administrators, science education specialists, and scientists, were established to write standards: One group focused on content standards, a second on teaching standards, and a third on assessment standards. When a draft document was finally produced after a long dialogue, it was sent out to thousands of teachers on all levels for their input before the final revision was made. The document we have today is the closest the United States has ever come to having a comprehensive national vision of science education.

Not only are the *National Science Education Standards* based on consensus, they are based on research. The Teaching, Professional Development, Assessment, Content, Program, and Systems Standards reflect current best practices derived from advances in education and the cognitive sciences. As new information becomes available about teaching and learning, revisions will undoubtedly be required. Right now, though, it is the best document we have to inform science teaching. After extensive critique and review lasting approximately two years, the *National Science Education Standards* were published in December 1995.

## A Vision for Science Education

The overarching goal of the *National Science Education Standards* is to provide for an education system that prepares a scientifically literate society. *Scientific literacy* is defined in the *Standards* as "the knowledge and understanding of scientific concepts and processes required for personal decision-making, participation in civic and cultural affairs, and economic productivity" (NRC 1996, 22). On the surface this statement seems to represent a reasonable goal for the majority of students. But, on reflection, one realizes that the achievement of scientific literacy is actually a lifelong process that is built on years of formal and informal instruction and experiences. In this sense, college science courses, especially for nonmajors, influence just a small fraction of a person's lifelong pursuit of scientific literacy. But when these courses build on a coherent and solid science background gained in K–12, the college science experience can be one that enriches people's lives. The powerful ideas that captivate and energize us as scientists can similarly change forever the way students view the world in which they live. However, what students learn and understand about the content and processes of science is greatly influenced by how they are taught.

The *National Science Education Standards*, which provide guidelines for how science should be taught, are based on four guiding principles, which may be paraphrased as follows (see NRC 1996, 9):

- Science is for all students.

- Students learn best by active participation in the learning process.

- Education should reflect the way that science is done.

- Improving science education requires a coordinated effort of *many* stakeholders to change the complex educational system. Stakeholders include teachers, supervisors and local communities, administrative personnel, policymakers, assessment specialists, curriculum developers, science educators, and more.

These four principles have important implications for science teaching—not only for grades K–12, but for the postsecondary years as well—and will be apparent in the narrative throughout this book. Three important terms from the field of educational research that are identified with these principles (and used throughout this book) are *constructivism*, *active learning*, and *inquiry approach*.

*Constructivism*. Science is most likely to be accessible to all students when students are engaged in the learning process and when they can see the importance of science in everyone's lives. Instructors need to recognize that students construct knowledge based on previous understanding and experience. This theory of learning is called *constructivism*, according to which students construct new understanding on an existing framework of knowledge (Lorsbach and Tobin 1993; Yager 1991; Bodner 1986). The basic premise of constructivism is that knowledge is built by the learner. That is, a person's knowledge of the world is developed over time through experiences that build upon what a person already knows. Thus each person's view of the world is unique. Modern constructivism, which can be traced back to Swiss psychologist Jean Piaget, is interpreted differently by different camps and as such is the subject of much debate. However, in spite of the lack of consensus regarding philosophical details, constructivist teaching models have evolved. These models redefine the teacher as a facilitator of learning rather than a disseminator of information and the student as an active participant in his or her own learning. Figure i.1 shows three of the fundamental differences between traditional and constructivist classrooms as presented by Brooks and Brooks (1993).

**Figure i.1**

*Fundamental Differences between Traditional and Constructivist Classrooms*

| Traditional Classrooms | Constructivist Classrooms |
|---|---|
| Students are viewed as "blank slates" onto which information is etched by the teacher. | Students are viewed as thinkers with emerging theories about the world. |
| Teachers generally behave in a didactic manner, disseminating information to students. | Teachers generally behave in an interactive manner, mediating the environment for students. |
| Assessment of student learning is viewed as separate from teaching and occurs almost entirely through testing. | Assessment of student learning is interwoven with teaching and occurs through teacher observations of students at work and through student exhibitions and portfolios. |

*Source:* Brooks, J. G., and M. G. Brooks. (1993) *The Case for Constructivist Classrooms.* Alexandria, VA: ASCD, 17.

A good constructivist science lesson, according to Saunders (1992) is one in which students are "thinking out loud, developing alternative explanations, interpreting data, participating in cognitive conflict, developing alternative hypotheses, designing further experiments to test alternative hypotheses, and selecting plausible hypotheses from among competing explanations" (140).

*Active Learning.* The principle of active participation indicates that students cannot be passive in the learning process. For example, Angelo (1990) has shown that students retain on average only 20 percent of what they hear in a lecture class. If learning is to occur in the classroom, science lessons must actively involve students both mentally and physically (Bybee 1993). Student activity, or *active learning*, needs to supplant the more traditional lecture approach to teaching.

*Inquiry Approach.* The inquiry approach to learning is based on the idea that science education should reflect the way that science itself is carried out. Inquiry can be defined both as a learning goal and an instructional strategy. As a learning goal, inquiry includes both the abilities to do scientific inquiry and a set of understandings about scientific inquiry. The abilities referred to are those that scientists commonly use to investigate their world. Understandings of scientific inquiry represent "how and why scientific knowledge changes in response to new evidence, logical analysis and modified explanations debated with a community of scientists" (NRC 2000, 21). One understanding, for example, might be the following:

> Scientific explanations must adhere to criteria such as: a proposed explanation must be logistically consistent; it must abide by the rules of evidence; it must be open to questions and possible modifications; and it must be based on historical and current scientific knowledge. (NRC 2000, 20)

Inquiry-based teaching and learning draw on instructional strategies that have the following essential features:

- Learners are engaged by scientifically oriented questions.
- Learners give priority to evidence, which allows them to develop and evaluate explanations that address scientifically oriented questions.
- Learners formulate explanations from evidence to address scientifically oriented problems.
- Learners evaluate their explanations in light of alternative explanations, particularly those reflecting scientific understanding.
- Learners communicate and justify their proposed explanations. (NRC 2000, 25)

As one might imagine, inquiry teaching and learning require students to be physically active and mentally engaged. Inquiry lessons can emerge in the classroom and in the laboratory.

The *National Science Education Standards* recognize that the responsibility for change requires a coordinated effort of many parties in order to be successful. In grades K–12, support for the teacher—and students—must come from administrative personnel (principals, superintendents), policymakers (school boards, government agencies), and the community (especially parents and businesses). At the undergraduate level, one can identify similar entities that will be necessary to foster widespread change in science teaching. Academic institutions will need, for example, to provide support for and recognition of quality teaching, as well as to seek ways to enhance communication with external communities and policymakers. Academic departments will need to encourage coordination among science programs, and scientists and professional societies will be important players in creating and sustaining change in science teaching at the postsecondary level.

# A Role for Higher Education

The goal of the *National Science Education Standards* is to achieve scientific literacy, and perhaps the best place to begin this process is in K–12 science classrooms, but this book supports the argument that higher education must be involved in changing the way that science is taught in K–16 and beyond. To support change in K–12, we must change the way that science is taught at the college level. We need to examine the way we teach *all* of our courses—the courses for science majors and health professionals, general science courses for nonscience majors, and preservice courses for future teachers. Each of these types of courses affects our citizenry—as our students become teachers, health professionals, manufacturers, parents, and so on. These are the people who will guide the future of society's endeavors, and they will base their decisions on their understanding of science and technology and its impact in their lives. Among the very important courses that we teach are those in which future K–12 teachers are enrolled; the students in these courses can exponentially expand the effect of a single college science course when they in turn teach young children in their most formative years. The preservice courses and the future teachers they serve will, in large part, determine the nation's comfort with, knowledge about, and interest in science.

A legitimate question to ask is, *Why is change important now*? There are many answers to that question. For example, the science that we communicate to our students is highly complex; decisions facing our citizenry need to be based on an understanding of science and technology; and, importantly, our students have changed. The life experiences of our students are not those that most teachers of science faced during their own formative years; in fact, the experiences of young adults today are significantly different from those of only ten years ago. Science and technology are shaping our evolving environment, and we must enable our students to make decisions and to participate effectively in the "new" world.

## References

Aldridge, B. G. (Ed.) (1996) *Scope, Sequence, and Coordination: A High School Framework for Science Education*. Arlington, VA: National Science Teachers Association.

American Association for the Advancement of Science (AAAS). Project 2061. (1989) *Science for All Americans*. New York: Oxford University Press.

American Association for the Advancement of Science (AAAS). Project 2061. (1993) *Benchmarks for Science Literacy*. New York: Oxford University Press.

Angelo, T. A. (1990) *Learning in the Classroom (Phase I)*. Berkeley, CA: University of California, Lawrence Hall of Science.

Bodner, G. M. (1986) Constructivism: A Theory of Knowledge. *Journal of Chemical Education* 63, 873-78.

Brooks, J. G., and Brooks, M. G. (1993) *The Case for Constructivist Classrooms.* Alexandria, VA: ASCD.

Bybee, R. (1993) An Instructional Model for Science Education. In *Developing Biological Literacy.* Colorado Spring, CO: Biological Sciences Curriculum Study.

Lorsbach, A., and Tobin, K. (1992) *Research Matters…to the Science Teacher: Constructivism as a Referent for Science Teaching.* National Association for Research in Science Teaching (NARST) Monograph 5, 21-27.

Moore, J. A. (1993) *Science as a Way of Knowing: The Foundations of Modern Biology.* Cambridge, MA: Harvard University Press.

National Commission on Excellence in Education (NCEE). (1983) *A Nation at Risk: The Imperative for Educational Reform.* Washington, DC: NCEE.

National Research Council (NRC). (1996) *National Science Education Standards.* Washington, DC: National Academy Press.

National Research Council (NRC). (2000) *Inquiry and the National Science Education Standards.* Washington, DC: National Academy Press.

National Science Teachers Association (NSTA). (1992) *Scope, Sequence and Coordination of Secondary School Science. Vol. I. The Content Core: A Guide for Curriculum Designers.* Arlington, VA: NSTA.

National Science Teachers Association (NSTA). (1993) *Scope, Sequence, and Coordination of Secondary School Science. Vol. II. Relevant Research.* Arlington, VA: NSTA.

Saunders, W. (1992) The Constructivist Perspective: Implications and Teaching Strategies for Science. *School Science and Mathematics* 92 (3): 136-41.

Yager, R. (1991) The Constructivist Learning Model: Toward Real Reform in Science Education. *The Science Teacher* 9, 53-57.

# Teaching Standards

William J. McIntosh

The vision portrayed by the *National Science Education Standards* (NRC 1996), referred to here as the *Standards*, recognizes our emerging understanding of the teaching-and-learning process. The vision is one in which effective teachers of science "create an environment in which they and the students work together as active learners" (NRC 1996, 28). This environment supports close student-teacher interaction and provides an opportunity to develop inquiry skills as a means of achieving scientific literacy. The research-supported vision of the *Standards* does not support the practice in institutions of higher education by which instructors lecture to groups of students who subsequently verify the concepts and principles in the laboratory. Although huge lecture halls and verification labs may be cost-effective, they are an ineffective means for students to learn concepts and practice science (Lord 1994; Leonard 1992). We know from cognitive psychology, for example, about the importance of providing opportunities for students to construct their own knowledge. A growing body of research supports the notion that misconceptions that students bring to a learning environment may interfere with the acquisition of new knowledge. Researchers have found, however, that specific teaching techniques can be used to change those misconceptions (Posner et al. 1982).

The *Standards* vision guides the discussion in this chapter on science teaching standards for the postsecondary level. The discussion centers on the importance of goal setting, designing experiences to meet students' needs, assessment, and collegiality. There is a strong recommendation that students be given opportunities to engage in meaningful scientific inquiry—to ask scientific questions, design experiments to collect evidence, and make critical interpretations of observations.

The ideas in this chapter and in the *Standards* challenge the traditional notion of the role of lectures in college science teaching. The authors whose work you are about to read—in the From the Field vignettes—are currently engaged in developing the practices about which they write. Their interpretations of the *Standards* offer us insights about how we can most effectively guide our students along the pathway to scientific literacy.

# Planning an Inquiry- Based Science Program

Teachers of science plan an inquiry-based program for their students. In doing this, teachers

● Develop a framework of yearlong and short-term goals for students.

● Select science content and adapt and design curricula to meet the interests, knowledge, understanding, abilities, and experiences of students.

● Select teaching and assessment strategies that support the development of student understanding and nurture a community of science learners.

● Work together as colleagues within and across disciplines and grade levels.

*Source:* National Research Council. (1996) *National Science Education Standards.* Washington, DC: National Academy Press, 30.

A central tenet of the *National Science Education Standards* is teaching science as inquiry, as discussed in the Introduction to this book. The effectiveness of inquiry as an instructional strategy has been well-documented across grade levels (Haury 1993). In a major study of college-level physics courses, Hake (In press) concluded that "the use of IE (Interactive Engagement) strategies can increase mechanics-course effectiveness well beyond that obtained with traditional methods." Hake's IE strategies closely align with a definition of "inquiry."

College instructors, like their K–12 counterparts, are addressing Teaching Standard A by looking more critically at the content of the courses they teach. In many cases higher education faculty are restructuring their courses to provide students with opportunities to pose and investigate researchable questions (Harker 1999; Weld, Rogers, and Heard 1999). They are doing so because, for all students, the development of scientific inquiry skills becomes an important component of their scientific literacy and problem-solving abilities. For science majors, effective scientific inquiry skills are a critical link between their K–12 science education and their careers or graduate research. It is especially important that students intending to major in the sciences improve on these abilities at the introductory college level in order to lay the foundation for upper-level course work.

Providing inquiry experiences in a college-level science course requires a different kind of planning than many higher education faculty members are used to. The planning starts with focusing on course goals that reflect both content and inquiry skills to be mastered. The nature of the experiences in which students will be engaged during the semester is defined; the experiences are chosen less by tradition than by appropriateness to course content and inquiry skills to be developed. In many cases students themselves choose questions that they are interested in and that they wish to investigate. College-level inquiry experiences, for example, can range from problem-based learning (Duch 1996), where students are given the opportunity to solve real world problems, to inquiry labs (Glasson and McKenzie 1998), where students design and conduct their own investigations.

Once goals are established, the logistics of inquiry teaching are addressed. In many large enrollment courses the lecture component generally focuses on content while the laboratory activities focus on science as an active process. In a smaller enrollment course, lecture and laboratory may be integrated to achieve similar goals. The best results seem to come about when students build competency with the scientific process in a stepwise fashion. In that scenario, an experimental sequence is planned that builds in complexity. Students begin by designing experiments to test relatively simple hypotheses and later to examine more complex ones. Laboratory instructors may lead discussions to help the students generate hypotheses that would be feasible to test within the constraints of budget, space, materials, personnel, and time. Students generate research questions, design experiments, collect data, and reach conclusions. They generally work collaboratively; in some cases, each member of a laboratory team is responsible for presenting an oral report on his or her aspect of the topic and illustrating his or her presentation with a poster.

Proper planning for inquiry-based learning is labor intensive for the course instructor, teaching assistants, and even the students. A great deal of effort must go into preparing experimental materials, critiquing protocols, supervising students, and mentoring teaching assistants to be good scientific inquiry coaches. However, students who design their own experiments report having a heightened sense of ownership, which in turn increases their motivation and interest in science (Stukus and Lennox 1995). Professors note that most students become highly involved in their research, teaching assistants are enthusiastic about teaching the course, and students' scientific inquiry and problem-solving skills improve.

## From the Field

### Changing Student Attitudes about Science through Inquiry-Based Learning

Gerald H. Krockover
Department of Earth and Atmospheric Sciences, School of Science
Department of Curriculum and Instruction, School of Education
Purdue University

I have been teaching sophomore/junior-level undergraduate science courses for nearly thirty years and have taken students with strong negative attitudes toward science and helped them to have positive learning experiences. A key to changing attitudes is to *engage* students in learning science.

I have used three techniques of engagement simultaneously: inquiry-based learning, relating science to career choices, and making important connections between science and other academic disciplines.

I begin my course with a presentation to students about the relevance of science and of this course to their career choices—whether that choice is education, liberal arts, engineering, or agriculture. Then we discuss ways of knowing and learning science. Members of the class work in collaborative science teams of five to six students each and focus their efforts on laboratory experiences for learning science. We stress the importance of using questions, developing hypotheses, conducting experiments, and collaborating with colleagues. We also stress connections between the sciences and mathematics, particularly by developing skills such as measuring and graphing and using technology and application software.

Maintaining contact among class members outside of class is essential to building a community of learners.

Thus, we keep in touch with each other via e-mail and the Internet so questions can be asked anytime not only of the instructor, but also of other students. I also have previous years' students online to serve as resources. And students participate in one field trip each term to emphasize the relation of course content to real problems. The trips became a course focus some years ago when a survey indicated that 95 percent of the students in my introductory course had not taken a field trip since elementary school.

As a result of these experiences, more than 1,800 of my students have modified their attitudes toward science. These students are confident that they can learn science content in an inquiry-based environment, relate science to their career choices, and make connections to other science and nonscience disciplines. They know that inquiry into authentic science questions has helped them to develop skills for lifelong learning. And, isn't this what college science teaching is all about?

## Guiding and Facilitating Learning

Teachers of science guide and facilitate learning. In doing this, teachers

- Focus and support inquiries while interacting with students.
- Orchestrate discourse among students about scientific ideas.
- Challenge students to accept and share responsibility for their own learning.
- Recognize and respond to student diversity and encourage all students to participate fully in science learning.

- Encourage and model the skills of scientific inquiry, as well as the curiosity, openness to new ideas and data, and skepticism that characterize science.

*Source:* National Research Council. (1996) *National Science Education Standards.* Washington, DC: National Academy Press, 32.

Teaching Standard B supports Teaching Standard A in that it addresses the nature of student-teacher interactions that support inquiry learning. Higher education faculty who meet this Standard guide and facilitate learning by recognizing the worth of all students and by communicating with them on a level that challenges their thinking and piques their curiosity. This can be accomplished in a small class by coordinating group discussions and relevant activities by which students carry out meaningful, inquiry-oriented tasks. Instructors skillfully weave content and inquiry skills as they model the kinds of thinking students are expected to exhibit. However, as one might imagine, this methodology poses a formidable (though not insurmountable) challenge in large enrollment classes.

In the laboratory, this inquiry-based instructional mode places both the instructor and the student in new roles (Glasson and McKenzie 1998). The students are actively engaged in the "minds-on" activity of designing experimental procedures. The instructor's role changes from being an authority figure who clarifies procedure and tells students whether they have the right answer to that of a facilitator: someone who interacts with the students to foster cognitive growth by asking them questions and guiding them in their inquiries.

In addition to guiding and supporting student inquiries, the facilitator encourages all students to participate fully in their learning while respecting and appreciating the qualities other students bring to the learning environment. One way to do this is to have the students work in small groups. Each group member has a role (e.g., leader, scribe, technician, or counselor) and shares in the responsibility of the learning process (Duch 1996). Role assignments change for each activity, thus preventing one person from dominating the group. This cooperative arrangement also helps students reflect on their strengths and weaknesses and encourages discussion among students. Within a group, students debate and justify ideas; conceptual understanding is reinforced as students try to persuade other group members of the validity of their ideas. To encourage group work, opportunities must be provided for the group members to meet and discuss the activity.

Group presentations are another way to encourage discussion among students. By presenting their work in an open forum, the students are given the opportunity to express their ideas and to debate their conceptual understanding of scientific ideas and methodologies. Students must not only persuade others that their work is meaningful, but also justify what they have done.

Higher education faculty who act as facilitators encourage students to accept and share responsibility for their own learning. They interact with students as much as possible by asking questions, facilitating discussion, and challenging students to clarify ideas and draw conclusions. This instructional environment provides the students with a glimpse of what characterizes science. Much of science involves the art of persuasion: using data and one's conceptual understanding to convince others of the soundness of one's ideas. Criticism and debate are important, as it is through these processes that validity and meaning are established.

Science is more than replicating the works of others. It requires an openness to investigation, and the path to be taken is not always clear at the outset. Students can understand this concept with the help of faculty who view themselves as facilitators of learning as well as purveyors of knowledge.

## From the Field

### Identifying and Using Learning Styles to Facilitate Instruction
Nannette Smith, Division of Natural, Behavioral, and Social Sciences
Bennett College

At the beginning of the 2000 fall semester, faculty members at our college who taught introductory biology gathered to plan for coming classes. One professor who had attended a Project Kaleidoscope (see page 135) summer workshop boldly announced what many of us had come to suspect: "The emperor has no clothes," she said. "We are teaching but our students are not learning. We must change." And change we did.

The introductory biology professors decided that knowledge of how each student learned would help us as instructors and empower our students to direct changes in our pedagogical techniques. In pursuit of those goals, I designed the following cooperative venture. I began the process by telling my students about the conversation and events that had led

to the need for change in the course structure. To teach them more effectively, I said, I needed their help. I went on to explain that each of them was a unique learner: They each had their own style of learning and their own learning experiences. To help them understand introductory biology, I would need to know about these learning styles and experiences. I challenged my students to accept and share responsibility for their own learning and "do science" at the same time. They accepted the challenge.

My first step was to educate both my students and myself about learning style theory. I used my course Web site at *Blackboard.com* to link to a site where each student would take the NC State Index of Learning Styles Questionnaire, *www2.ncsu.edu/unity/ lockers/users/f/felder/public/ILSdir/*

*ilsweb.html*, online and free-of-charge. The site also provided us with two articles that contained in-depth explanations of theories of learning styles and their practical applications. As the professor, I was responsible for providing copies of the articles. After reading the articles, each student formed a hypothesis as to what the inventory would show her or his learning style to be. Students then took the inventory and printed out two copies of their results. One copy was placed in each student's file; the other was kept by the student. When all students had completed the assignment, they brought the results to class. Each student presented her or his results to the class and compared the inventory results to her or his hypothesis. Much discussion ensued. Some agreed with the inventory results and some did not. The students drew up conclusions as to how they would alter note taking, classroom questioning, and study habits as a result of the inventory. Each student's hypothesis, process, results, and discussion were combined in a written report that was turned in to me. We also discussed the changes that I would have to make in my classroom presentation in order to support the learning styles of all students.

In the true spirit of scientific research we quantified our results. We compiled total class results using Microsoft Excel to create tables and graphs of our data. Students unfamiliar with Excel were responsible for arranging tutorials with me or with classmates. Numerous e-mails followed as students explored how to graph and what graphs were most illustrative of the data gathered. The resulting graphs were included in their written reports.

The introduction of learning style theory, the cooperative investigation, and the inventory results significantly altered the nature of student-professor interaction in my class. Whereas students had formerly asked me simply to repeat material or to provide further explanation when they were confused, now their requests became much more specific and helpful to me. Students now asked if they could be provided with graphics, models, or sequential explanations to help them understand a concept. Oftentimes they cited their learning styles as justifications for their requests. Additionally, students began to help their peers. More important to me were their suggestions as to how I might alter a learning experience in order to address the styles of classmates.

I was thrilled when non-biology majors and students from other colleges began to access my site to take the inventory. However, my greatest reward came when several of my students commented, "Why haven't we been told this before? Everyone needs to know this!" Now, one hopes, everyone will.

# Linking Assessing, Learning, and Teaching

Teachers of science engage in ongoing assessment of their teaching and of student learning. In doing this, teachers

● Use multiple methods and systematically gather data about student understanding and ability.

● Analyze assessment data to guide teaching.

● Guide students in self-assessment.

● Use student data, observations of teaching, and interactions with colleagues to reflect on and improve teaching practice.

● Use student data, observations of teaching, and interactions with colleagues to report student achievement and opportunities to learn to students, teachers, parents, policy makers, and the general public.

*Source:* National Research Council. (1996) *National Science Education Standards.* Washington, DC: National Academy Press, 37-38.

Teaching Standard C invites us to consider the important links among assessment, learning, and teaching. A comprehensive assessment of students informs the instructor about student progress toward course goals (learning) and also about the efficacy of the instructor's teaching. Data collected both formally and informally can be used to identify students' prior knowledge and experiences, evaluate their understanding of important concepts being taught, and measure the extent to which students meet course goals. Instructors can use this information, in turn, to make decisions about instruction and subsequent student learning. Quality assessment also can provide students with feedback about their own progress toward course goals.

In general, a good assessment plan is systematic, multiple in form, and at times nonevaluative (see Chapter 3). This Standard distinguishes between the broader terms *evaluation* and *assessment.* In the case of evaluation, students are administered testing instruments for the purpose of assigning grades; assessment of student learning has the dual purpose of providing student feedback and improving the program. This means that assessments should be carefully planned so the feedback is useful for both student and instructor.

Instructors engage in *multiple forms of assessment* when they collect many different kinds of information about their students. Multiple forms of assessment provide a clearer picture of what students know and are able to do. In the classroom, traditional tests, quizzes, and laboratory work are supplemented or replaced by student presentations and products. Students can be asked to demonstrate knowledge application by conducting and presenting the result of long-term scientific investigations; instructors may also design performance assessments, tasks that require application of skills and knowledge to answer open-ended questions. In other words, students are being asked to think and reason as scientists do.

Other types of assessments include querying students about their knowledge of a subject before instruction in the subject begins. Because students construct new understandings on a preconceived base of knowledge (as discussed in the Introduction), it is important to identify student misconceptions so that instruction can be planned to counter them. Post-course queries can then be used to gauge progress toward goals established by the instructor.

Many instructors, before a major evaluation, engage in frequent and ongoing informal questioning of students that serves to inform both the student and teacher about the extent to which the class is reaching course objectives. This information can help students and instructors to identify difficult concepts that will require additional explanation and study. Faculty should also encourage students to use frequent self-assessments to guide their studies.

The design of assessment tools is discussed in more detail in Chapter 3. In the present chapter, the emphasis is on using carefully planned and frequent assessments in multiple forms to improve teaching. Such assessment data, in conjunction with classroom observations and professional interaction with colleagues, give instructors the information necessary to improve teaching—and student learning. These data also allow us to evaluate student achievement (i.e., assign grades) and to share with others the success of our programs.

In today's college classrooms, faculty are allowing students to demonstrate their understanding in a variety of ways. Multiple assessments access different sensory capabilities and as such are of particular importance to increasing the diversity of successful students in science fields—including, especially, the capable student who has recognized learning challenges and students who come to class with a wide range of cultural experiences (see also Chapter 5, Program Standard E). In short, multiple assessments provide valuable information about students that if used appropriately can guide both teaching and learning.

## From the Field

### Assessment Techniques to Guide Teaching in Courses with Large Enrollments

Harry L. Shipman, Department of Physical Sciences
University of Delaware

I have been teaching high-enrollment physics and astronomy courses composed of approximately five hundred students for years—and for many years my assessment of student work involved an hour exam or two and a final examination. Multiple-choice questions and fill-in-the-bubble answers to artificially concise problems led my students to rely on recall of material covered in class and in the text. I was unhappy with what my students "learned"—and what my assessments indicated were important for them in solving "real-life" problems. Over the past few years I have begun to develop a variety of different assessment techniques. Some of these techniques permit me to assess student work before the traditional hour or final examination in a more systematic way than "reading the faces" in class, or reacting to that uncomfortable atmosphere when students put down their notebooks, stare off into space, and otherwise indicate their disengagement. And importantly, a variety of types of assessments permits students to show what they know in different ways.

A guiding principle that I use in designing assessments is to make the assessments match the real world as much as possible. Given the constraints of a large class and the responsibility to assess the performance of individual students as well as teams, let me describe some of the ways I now assess my students.

Often during the semester, I stop just before the end of the class period and ask each person to take one minute to answer a simple question: "As you think about the topics we have studied so far in this class, what is the most confusing concept or phenomenon that we have studied?" I then collect the answers and read them. I try to clear up the most common points of confusion during the next class peri-

od. The question you ask can vary; Frederick Mosteller of Harvard asks students to identify the "muddiest point" in his statistics classes. This "minute paper" is one of the easiest of the new assessments to adopt. It has no doubt been reinvented hundreds of times; I first ran into it when I gave a seminar for the Scholar-Leader Enrichment Program at the University of Oklahoma. It was the minute paper concept that started me thinking about the broader definitions of assessment, in particular finding out where my students were conceptually before giving them an examination. [For more on the One-Minute Paper, see pages 156–157. –Ed.]

During class, my students will often work in groups. Every element of group work produces some kind of tangible product—a worksheet, observations, evaluations, or something—that counts for course credit. Were I to teach a small class, I might be able to grade these group products and give different groups different amounts of credit for what they do. Since my own classes are large, I simply give everyone participation credit for what they did. When it's practical, I'll ask students to evaluate each member's contribution to the group—at midsemester for diagnostic purposes and at the end of the semester for part of their grade. (It is important to do the midsemester evaluation so that students who aren't contributing to their groups can be given a chance to change their ways.) Instructors should make sure that raw student evaluations of their peers do not play too large a role in overall course grades.

Halfway through the semester, my students in the laboratory are challenged to come up with their own investigation or Big Project. Students spend three weeks on their Big Projects, and then present them in a meeting of their laboratory section. They have tested detergents, mailed potato chips across the country in different packages, built toothpick bridges, and dropped carefully packaged eggs off seventeen-story buildings. Their presentations are evaluated on content and on communication skills (which need to be overtly taught). We have found that it helps to have someone other than their regular laboratory instructors judge their work. The highest-ranked projects are then presented to the campus community, along with projects from a biology course, in a poster session.

On the final examination, I include a variety of recall, reasoning, and higher-order thinking questions. There are the usual multiple-choice and true/false questions; in addition, students must draw conclusions from graphs and tables. I present them with concept maps, and they have to fill in the blank spaces. (If I had a small class, students would draw the concept maps themselves and I'd evaluate them.) Sometimes they have to write essays, though I have found that giving them some questions ahead of time and having a few of those actually appear on the examination produces much more coherent answers.

Why use such a variety of assessments? What was wrong with the old way, with an hour exam or two and a final? The *National Science Education Standards* and other reform documents such as Project 2061 give a lot of reasons why multiple assessments are preferred. Two stick out in my mind as being particularly important. First, I want to make assessments that match the real world as much as possible in order to prepare students to tackle and solve problems they meet outside of class and in their later lives. Second, I learned some years ago that a teacher's tests define the real goal of the course for the students. If your syllabus contains a lot of words about developing critical thinking skills and your tests ask "What is X?", your students will quickly catch on that what really counts is memorizing vocabulary.

In summary, no single form of assessment provides me with an adequate picture of what students know—and understanding what students know is the key to better teaching.

## Designing and Managing the Learning Environment

Teachers of science design and manage learning environments that provide students with the time, space, and resources needed for learning science. In doing this, teachers

- Structure the time available so that students are able to engage in extended investigations.
- Create a setting for student work that is flexible and supportive of science inquiry.
- Ensure a safe working environment.
- Make the available science tools, materials, media, and technological resources accessible to students.
- Identify and use resources outside the school.
- Engage students in designing the learning environment.

*Source:* National Research Council. (1996) *National Science Education Standards.* Washington, DC: National Academy Press, 43.

Inquiry-based learning experiences often present logistical difficulties early in the assignment, as students form small groups and choose, within limits, topics to study. Many groups, for example, quickly face difficulties they are unprepared to resolve. Most problems stem from their ignorance of what physical quantity should be measured, or they may want to measure quantities their tools aren't designed to measure. However, designing a laboratory or field study forces the students to look more closely at their experimental design and the capabilities of their tools.

For faculty who are committed to providing inquiry experiences, new challenges of time, space, and resources arise.

*Time:* Often the hours of class time allocated to the projects are inadequate, and students (and instructors) must be willing, if necessary, to meet for additional hours to complete a project. Teacher oversight and guidance is necessary, and instructors spend considerable energy to keep the process going.

*Space:* Classrooms and laboratories are typically limited in their flexibility and tend to foster activities that are singular of purpose. In addition, space problems may arise when groups of students simultaneously differ in their experimental method and the materials they subsequently require.

*Resources:* Open-ended inquiry also poses the problem of finding available resources for students who choose their own topic to investigate. In some cases, the scope of an investigation is limited by the equipment available. Materials, too, may not be available or may need to be acquired on short notice. This is in contrast to the direct instruction approach, which essentially requires chalk and a blackboard; for laboratory experiments designed to verify pre-specified concepts, a semester's worth of laboratory materials can usually be ordered in advance, since student behaviors and experiences are predictable. Inquiry-based learning produces unanticipated learning excitement—and unanticipated challenges of course management!

Limited time, space, and resources can pose unique barriers to the acquisition of scientific inquiry skills. Despite those management challenges, however, college science teachers believe the time spent to be worthwhile, as the following vignette demonstrates.

## From the Field

### Developing Experimental Design Skills of Students
Patrick Gleeson, Department of Physics and Pre-Engineering
Delaware State University

My experiences with inquiry teaching have met with mixed results. The first time I had students conduct their own investigations, I was pleasantly surprised. Small working groups were formed and everything proceeded smoothly. Students actually recommended including the inquiry-based project into the course structure, even though it required more work on their part.

My second experience was different. Of six working groups, only one group selected its own topic; the remaining groups chose from the instructor's list. The group that selected its own topic chose to investigate the magnetization of different materials. They made their own electromagnets, studied the characteristics of the magnetic fields produced, and then looked at how these fields were influenced by the introduction of different materials

into the core of the electromagnet. Their study brought them past the brief discussion of magnetic materials they received in their course work, and well beyond any experiences they had had in making and measuring magnetic fields. They developed ingenious ways to identify and quantify the changes in magnetic field.

Another (single-person) group looked at the effects of salinity concentrations on the conductivity of liquids. He planned and conducted his experiment with great care, using equipment borrowed from another department. Unfortunately, he exercised no flexibility in reshaping his experimental plan when his preliminary measurements led to unexpected results. Instead, he approached this project like any other laboratory exercise and missed the opportunity to pursue questions raised by his own work.

Two of the other groups had difficulty organizing any form of systematic investigation. With the instructor's assistance, both groups completed satisfactory investigations. The remaining two groups performed poorly. Both groups reached inaccurate conclusions by incorrectly interpreting poorly collected data. Their experimental design was feasible, but they ignored the need to keep some of the parameters constant while investigating the effect of changing others.

Why were these experiences so different? The second class seemed to be intimidated by the concept of an open-ended project. Most of this class did not share the enthusiasm of the previous class, and I found myself coaxing and sometimes pulling them along just to keep them working. Without strong intervention, I'm certain that two-thirds of these students would have foundered. The quality of the work completed by this class was, at best, mixed. What clearly emerged from these experiences is the inability of traditional laboratory exercises to prepare students for the study of real world physical systems. Without the guidelines provided in the traditional laboratory, students had great difficulty deciding what to measure and how to measure it. Even after deciding what parameter should be measured, they selected the wrong power supplies, chose multimeters when they really needed oscilloscopes, and made many other poor instrumentation choices.

How could this happen? Our students receive extensive instruction on the content of our discipline. They have access to modern instrumentation and are thoroughly instructed on the rules of its use. Until my course, however, students had had no opportunity to use this knowledge to formulate and execute a study of their own. We have been so busy teaching science *to* our students that we failed to let them *do* science. We have failed to challenge our students with any relevant scientific problems to solve. When students are given the opportunity to study practical problems, the application of laboratory techniques and instrumentation that give unpredictable results force the students to think about their analytic and laboratory work at a level not encountered anywhere else in the curriculum. Even the undergraduates invited to work in our research laboratories are seldom given the opportunity to influence the direction of a research project.

From a practical perspective, I could not have predicted which students would most eagerly engage in the open-ended project. I observed one of our best students reduce the entire project to a thoughtless laboratory exercise, while some marginal students became very enthusiastic and imaginative investigators. Every group needed some guidance, but those groups that selected their own projects worked more independently than those that chose one from my list. While there may be several reasons for this, it's tempting to attribute the added enthusiasm to "ownership" in the project. From my experience, inquiry-based learning is much more effective when a group selects its own project for study.

Finally, for most students the prospect of presenting an oral report of their work to their peers was more frightening than receiving a poor grade. Knowing they would have to defend their techniques and conclusions imposed an unusual sense of thoroughness and honesty on their work. The presentations provided an excellent opportunity to practice communication skills; more importantly, they clearly illustrated the diverse ways that students tackled unique problems.

## Building Learning Communities

Teachers of science develop communities of science learners that reflect the intellectual rigor of scientific inquiry and the attitudes and social values conducive to science learning. In doing this, teachers

- Display and demand respect for the diverse ideas, skills, and experiences of all students.
- Enable students to have a significant voice in decisions about the content and context of their work and require students to take responsibility for the learning of all members of the community.
- Nurture collaboration among students.
- Structure and facilitate ongoing formal and informal discussion based on a shared understanding of rules of scientific discourse.
- Model and emphasize the skills, attitudes, and values of scientific inquiry.

*Source:* National Research Council. (1996) *National Science Education Standards.* Washington, DC: National Academy Press, 45-46.

Two observations about life relate directly to Teaching Standard E: (1) Every individual is unique, and (2) living organisms interact with each other and with the environment.

Teaching Standard E emphasizes diversity and respect for individuals. In this context, educators at all levels should help students to identify and nurture their unique strengths as well as help them to identify weaknesses and work to overcome them. Of course, some might raise the following questions: If each individual has unique talents, then why should we encourage collaborations and learning communities? Shouldn't we encourage each unique student to learn independently? Is Teaching Standard E contradictory in this regard? The observation that living organisms interact with each other and their environment serves as an analogy for the learning community environment. Many different individuals working together can provide unique insights and perspectives that extend beyond the thoughts of any single individual. Uniqueness of individuals does not disrupt the learning community environment—it enhances it. The instructor, who is the key to establishing the learning environment, should recognize individuality and encourage students to use collectively their unique talents to learn science.

The ideal college science classroom should be an interactive setting where students feel comfortable with the instructor, the teaching assistants, and with each other. All individuals should be readily exchanging thoughts about meaningful, interesting scientific concepts and issues, without fear of "saying something stupid." The prevailing attitude should be one of respect for different ideas and a shared eagerness to learn more about the scientific topic. The most effective learning happens when students view the course experience as something special, of which they are all a part. If the instructor can generate and support such interactions and excitement, the stage is set for learning.

Although student input is important in the planning process, the instructor must take the lead. Asking students to help plan a course that they have never had seems unreasonable. The instructor should propose a framework, and then respond to appropriate input and modifications from students. Overall, such shared planning can provide the basis for the desired learning community. The following vignette illustrates many features that may be incorporated into a program to foster community.

## From the Field

### Creating a Learning Community in Introductory Biology

Marvin Druger, Department of Biological Sciences/Science Education
Syracuse University

I have taught introductory college biology for more than forty-four years and have evolved mechanisms that seem relevant to Teaching Standard E. The introductory biology course is a two-semester course that serves both science majors and nonmajors. The course involves lectures, recitations, and laboratory work. I try to establish a learning environment where every individual feels important and comfortable. I constantly remind my students that each of them is unique and that my role as a teacher is to help them maximize their individual talents and learning. I try to emphasize that my role is "to help you learn" and not to "fail you" or "get you" or separate the "geniuses from the dummies." I would be delighted if my students all learned 90 percent of the content and received an A in the course. Indeed, I sometimes tell students some of the exact questions that will be on the exam (in modified form), and old exams are available for their review. Why should students have to guess what I think is important for them to know? Why can't I tell them, and then expect them to learn that content and do well on the exams and be rewarded? I often repeat questions (in modified form) that students had wrong on a previous exam, as bonus questions. Once students feel comfortable in class and have a good rapport with the teacher, they have an environment in which they feel free to learn as much as they can.

My course contains many diverse features that are designed to make it a unique set of experiences that extend beyond mere acquisition of content knowledge. I want my course to have a positive impact on students many years after they have completed it. Special course features include the following:

- *Bionews and Bioviews* (a weekly course newsletter)
- Bio-Lunches (lunch dates with small groups of students in the dining halls)
- *Bio-Answer Show* (a closed-circuit TV program during which I review answers to the exam and give away prizes in a drawing of student names from the bio-fishbowl)
- Support for taking examinations (e.g., review sessions before major exams; having old exams available in the library; bonus questions on exams [students love extra credit])
- Benefit-of-the-doubt credit (credit for students who attend extra sessions and lectures on campus—e.g., an optional Wednesday session [to enhance the content for the more interested students]; the Frontiers of Science Lecture Series [science faculty tell about their re-

search]; Pathways to Knowledge Lecture Series [Ph.D. students present their research])
- Bio-Creativity contest (students create something about "life"—e.g., a poster, poem, essay, or model—and receive ten, five, or two points toward their final grade)
- A cooperative research project on plants
- A pig dissection, followed by a "pig interview" with a teaching assistant (nobody leaves the course without a one-on-one consultation with an instructor)
- Support for self-learning (e.g., audiotapes, CD-ROMs, demonstrations, models, videotapes)

These special features are intended to create a sense of excitement, so that students feel proud that they are a part of a set of unusual experiences designed to help them learn.

The laboratory is open seven days a week (including nights) to help students learn to manage their own time. A teaching assistant is always available, and students are encouraged to drop in to do laboratory work whenever they have the urge to do so. I want students to talk to each other and to teaching assistants, so the laboratory becomes a social environment where students learn from and interact with each other.

Although Teaching Standard E emphasizes the importance of nurturing collaboration among students, this may not be desirable for all individuals. Some students do not want to work in groups on a research project or a dissection. Should we force such students to work collaboratively? I think not. I try to encourage students to collaborate, but I respect the individuality of a student who wants to work alone, and such students can do so in my course.

In general, we should try to use our course experience to meet the needs and interests of students, be partners in the learning process, create an exciting learning environment, and help students to learn in such a way that the course is still having an impact on their lives twenty years later.

## Participating in Program Development

Teachers of science actively participate in the ongoing planning and development of the school science program. In doing this, teachers

● Plan and develop the school science program.

● Participate in decisions concerning the allocation of time and other resources to the science program.

● Participate fully in planning and implementing professional growth and development strategies for themselves and their colleagues.

*Source:* National Research Council. (1996) *National Science Education Standards.* Washington, DC: National Academy Press, 51.

Although Teaching Standard F is written for K–12 science teachers and addresses school science programs, it is pertinent to university and college teachers of science with slight adaptation. The Standard notes that K–12 science teachers should participate in program development in their departments and their institutions with careful attention to Teaching Standards A through E. To accomplish this on the college level, science teaching faculty should first understand the conceptual basis of national, state, and local initiatives of science reform; then, they can put the initiatives into practice. Thus, in all teaching, science faculty should strive to

● connect science concepts to real world issues relevant to the lives of students;

● actively engage in ongoing program development and assessment to meet changing needs of student populations;

● collaborate and network with other colleges, departments, school systems, informal science centers, and the public and private sectors during the process of program development; and

● foster a vision of scientific literacy that encourages practical knowledge of the nature of science, developing habits of mind consistent with positive scientific perspectives and attitudes, stressing skepticism and critical thinking, teaching for conceptual understanding of seminal linking themes and theories among the sciences, embedding science in cultural and historical contexts, and providing opportunities for students to generate their own meaningful questions and design approaches to investigate real world issues.

Teaching Standard F may be considered somewhat of a double-edged sword. On one edge, the Standard appears to be an obvious statement that the K–16 (and beyond) teacher of science would assent to: that science teachers should exercise both their academic freedom and content knowledge expertise in the planning and development of new science curricula. The need to update existing science courses or formulate new courses based on current research relevant to a field of study has been a historic and continuing concern in college science teaching. Despite the perennial "updating" of course material, however, very little else seems to change in higher education in terms of the design and delivery of most foundation science courses. But the opposite edge of the sword suggests another way to think about such program development: from a bottom-up, interdisciplinary, global perspective. Change for the sake of change is probably benign at best, and foolhardy at worst. But change in response to recent research about how students construct knowledge, seek relationships in nature, and demand to know how science is relevant is imperative; and change in response to how students will negotiate their lives through a maze of real world issues and problems is, at the very least, prudent.

College and university science teachers and science educators are in a unique position to engage in curriculum reform that reflects national K–12 science education reform initiatives. To make an informed decision about the extent to which one wishes to participate in program development and restructuring, higher education science faculty must make a concerted effort to understand the visions of future scientific literacy, the thematic strands that permeate science, the ways in which students learn science, science's connections to other disciplines, multiple forms of assessment, and ongoing program evaluation efforts. Rethinking logistical arrangements of course delivery (e.g., scheduling that allows more time to pursue small-group collaborative research projects, working with mentors, shadowing professional scientists, developing partnerships with industry) is now a viable option. Although being involved in such rethinking is not a simple undertaking, it does represent a full expression of academic freedom and empowerment by university science instructors who want to affect their students' lives through a systemic approach to program development.

# From the Field

### Aligning Courses for Standards-Based Teaching
Dana L. Zeidler, College of Education
University of South Florida

Recently I was offered an opportunity to develop an introductory science course that would bear little resemblance to traditional courses for nonscience majors (which could count as an elective for science majors). A colleague in the Department of Biology at the University of South Florida where I teach asked if I might be interested (as a science educator) in ferreting out seed money to create a new course for the College of Arts and Sciences that was consistent with the common themes of national science reform initiatives.

Here was an opportunity not only to talk the (reform) talk but to walk the walk. As we discussed ideas a bit further, we identified a small agency through NASA/NOVA that would fund teams of faculty and administrators who wished to either modify an existing course or develop a new course that echoed the science reform initiatives (and incorporate some of the goals of NASA as well). It was preferable that such teams consist of a science or mathematics educator, a faculty member from a science discipline, and an administrator (e.g., faculty chair, associate dean, dean) from the university. We found our third partner in the form of an associate dean—also a member of the Department of Biology—who shared our interests in developing alternative courses that would better engage students in scientific thinking and discourse. Putting together such a collaborative team would prove to be imperative in weakening the rigid college boundaries and personal niches that traditionally exist within many university structures.

As our team shared knowledge of local, university, state, and national resources with each other, a link between an aging population and the space industry seemed natural to us. A preservice science education/general education course (taught out of a science department) that incorporated the elements of long-distance and longtime space travel, the use of aging astronauts in space research, and natural developmental and aging processes on Earth would, we concluded, have broad appeal and relevance to future teachers and citizens in Florida. And so our team members collaborated with each other, sharing content and pedagogical knowledge, and directed our efforts in the "selling" of this new course to the College of Arts and Sciences and the College of Education by using the national science reform documents as our guide. Our emphasis was on systems, models, constancy and change, and scale. The decision by NASA to send John Glenn back into space reaffirmed a long neglected idea for us—that age has value. Furthermore, the need for long-term endurance as we travel increasingly farther from Earth emphasizes the need to understand better the aging process. From this premise, a new course emerged: Space Age Biology.

We think that a course about the concerns and challenges of long-range space travel and the normal problems of a senior citizen functioning in space will enable us to focus also on the biology of aging on Earth. Consider the following examples:

- A potentially lackluster unit on energy flow, metabolism, and respiration is enlivened and enriched by discussions and activities that focus on survival in a self-contained closed system such as the International Space Station. We include an activity in which our heterogeneous students (mixed ages) create a plan for spending a week together in a room the size of the living quarters of the space station. Respiration, nutrition, excretion, exercise, and entertainment have to be managed, along with the unique needs of seniors.

- A discussion on the genetic limits of aging and the relationship of life span to metabolic rates is related to time limitations in space, modifications of metabolic rates, and the possible use of technology to extend life spans. Birds are famously long-lived yet have very high metabolic rates—a seemingly inherent contradiction. Why? Scientists have recently re-implanted

modified drosophila genes in fruit flies and managed to increase their lives by 40 percent! Students may design small-scale projects involving metabolism and/or genetics for future consideration by NASA scientists.

Examples such as these illustrate how students are engaged through topics meaningful to their lives. NASA data will stimulate their thinking, and through their own designed experiences and problem-solving challenges, they will, one hopes, apply the science concepts in ways that will improve and extend their lives.

Our efforts to seek new pathways to participate in program development have stemmed from the common pedagogical themes that are woven throughout all of the national science reform initiatives. Our team will continue to develop and implement several of the most pervasive themes that recommend fundamental changes in how we teach science. The most important of these themes include the practice of:

- constructivist-based understanding of learning;
- hands-on/minds-on, active, problem-solving investigations;
- emphasizing the interdisciplinary connections of science and the history and nature of science;
- relating science to the students' world;
- consideration of personal, social, and ethical issues;
- focusing on fewer science topics in more depth;

- full integration of appropriate technology in instruction;
- teachers being facilitators of learning and learners as well;
- cooperative learning situations;
- science as argument and explanations; and
- multiple (alternative, in conjunction with traditional) forms of assessment.

If we believe that higher education must ensure scientific literacy for all of our students, then elements of these themes must be present throughout our curriculum. And if we are willing to invest our energies in collaborative program development, then the paths we choose are more likely to be responsive to the national reform initiatives and the changing needs of our students.

## References

Duch, B. (1996) Problem-Based Learning in Physics: The Power of Students Teaching Students. *Journal of College Science Teaching* 25 (5), 326-29.

Glasson, G. E., and McKenzie, W. L. (1998) Investigative Learning in Undergraduate Freshman Biology Laboratories. *Journal of College Science Teaching* 27 (5), 189-93.

Hake, R. R. (In press) Interactive Engagement vs. Traditional Methods: A Six Thousand Student Survey of Mechanics Test Data for Introductory Physics Courses. *American Journal of Physics.*

Harker, A. R. (1999) Full Application of the Scientific Method in an Undergraduate Teaching Laboratory. *Journal of College Science Teaching* 29 (2), 97-100.

Haury, D. L. (1993). Teaching Science Through Inquiry. ERIC CSMEE *Digest* (March, ED 359-048).

Leonard, W. (1992) Lecturing Using Inquiry. Paper to the Society for College Science Teachers. National Science Teachers National Convention, Boston.

Lord, T. (1994) Using Constructivism to Enhance Student Learning in College Biology. *Journal of College Science Teaching* 23, 346-48.

National Research Council (NRC). (1996) *National Science Education Standards*. Washington, DC: National Academy Press.

Posner, G. L., Strike, K. A., Hewson, P. W., and Gertzog, W. A. (1982) Accommodation of a Scientific Conception. *Science Education* 66, 211-27.

Stukus, P., and Lennox, J. (1995) Use of an Investigative Semester-Length Laboratory Project in an Introductory Microbiology Course. *Journal of College Science Teaching* 25, 135-39.

Weld, J., Rogers, C., and Heard, S. (1999) Semester-Length Field Investigations in Undergraduate Animal Behavior and Ecology Courses. *Journal of College Science Teaching* 28 (5), 326-29.

## Bibliography

Abdi, S. W. (1997) Multicultural Teaching Tips. *The Science Teacher* 64, 34-7.

Anderson, H. O. (1994) Teaching Toward 2000: Examining Science Education Reform. *The Science Teacher* 61, 49-53.

Beck, M. T., and Kauffman, G. B. (1994) Science Methodology and Ethics in University Education. *Journal of Chemical Education* 71, 922-24.

Berger, M. (1994) What Makes a Community Click? *Instructor* 104, 45-6.

Boujaounde, S. (1995) Demonstrating the Nature of Science. *The Science Teacher* 62, 46-9.

Brunkhorst, B. J. (1996) Assessing Student Learning in Undergraduate Geology Courses by Correlating Assessment with What We Want to Teach. *Journal of Geoscience Education* 44, 373-78.

Cavalli-Sforza, V., Weiner, A. W., and Lesgold, A. M. (1994) Software Support for Students: Engaging in Scientific Activity and Scientific Controversy. *Science Education* 78, 577-99.

Driver, R. et al. (1994) Constructing Scientific Knowledge in the Classroom. *Educational Researcher* 23, 5-12, 21-3.

Duggan, S., Johnson, P., and Gott, R. (1996) A Critical Point in Investigative Work: Defining Variables. *Journal of Research in Science Teaching* 33, 461-74.

Fort, D. C. (1995) Top Federal Science Agencies Join Other Reformers to Focus on the Vital Undergraduate Years. *Journal of College Science Teaching* 25, 26-31.

Giordan, A. (1995) New Models for the Learning Process: Beyond Constructivism? *Prospectus* 25, 101-18.

Haney, J. J. (1996) Teacher Beliefs and Intentions Regarding the Implementation of Science Education Reform Standards. *Journal of Research in Science Teaching* 33, 971-93.

Hartman, E. M., Jr., and Dubowsky, N. (1995) The Nature and Process of Science: A Goal Focused Approach to Teaching Science Literacy. *Journal of College Science Teaching* 25, 92+.

Havas, E., and Lucas, J. (1994) Modeling Diversity in the Classroom. *Equity and Excellence in Education* 27, 43-7.

Herwitz, S., and Guerra, M. (1996) Perspectives, Partnerships and Values in Science Education: A University and Public Elementary School Collaboration. *Science Education* 80, 201-27.

Hobson, A. (1994) Incorporating Scientific Methodology into Introductory Science Courses. *Instructional Science* 22(6), 413-22.

Howe, K. R. (1994) Standards, Assessment and Equality of Educational Opportunity. *Educational Research* 23, 27-33.

Hurtado, S. (1996) How Diversity Affects Teaching and Learning. *Educational Research* 77, 27-9.

Kilcrease, K. (1995) Creating a Sense of Community for Adult Students. *Adult Learner* 6, 7.

Lewis, A. C. (1997) Changing Assessment, Changing Curriculum. *Educational Digest* 62, 13-17.

Linder, C. J. (1993) A Challenge to Conceptual Change. *Science Education* 77, 293-300.

Lovely, C. et al. (1996) Student Abstracts, Scientific Method, and Critical Thinking. *College Student Journal* 30, 516-18.

Mason, L. (1999) Science and Math Curriculum in the 21st Century. *Journal of College Science Teaching* 25, 346-51.

MacPhee, D. et al. (1994) Strategies for Infusing Curricula with a Multicultural Perspective. *Innovative Higher Education* 18, 289-309.

McIntosh, W. J. (1996) Assessment in Higher Education. *Journal of College Science Teaching* 25, 52-3.

Micikas, L. B. (1995) Teaching about the Nature and Process of Science— Ask "Why" Before Considering "How." *Journal of College Science Teaching* 25, 99-101.

Nitko, A. J. (1995) Is Curriculum a Reasonable Basis for Assessment Reform? *Educational Measurements* 14, 5-10+.

Norris, N. (1995) Learning to Live with Scientific Expertise: Toward a Theory of Intellectual Communalism for Guiding Science Teaching. *Science Education* 79, 201-27.

Richmond, G., and Striley, J. (1996) Making Meaning in Classrooms: Social Processes in Small Group Discourse and Scientific Knowledge Building. *Journal of Research in Science Teaching* 33, 839-58.

Shymansky, J., Enger, S., Chidsey, J., Yore, L., Jorgensen, M., Henriques, L., and Wolfe, E. (1997) Performance Assessment in Science as a Tool to Enhance the Picture of Student Learning. *School Science and Mathematics* 97, 172-83.

Terry, C. (1996) Not Quite a Workshop, No Longer a Lecture: Implementing Inquiry-based Physics in the Community College. *Journal of College Science Teaching* 25, 297-99.

Trice, A. D. (1996, Summer) Men's College Classroom Environments. *Education* 116, 572-77.

Walberg, H. J. et al. (1994) Assessment Reform: Challenges and Opportunities. *Phi Delta Kappa Fastbacks* #377, 7-41.

Wilde, S., and Webb, T. (1996) Creating a Classroom Community. *Learning* 24, 85-6.

Williams, K. A., and Cavallo, A. M. L. (1995) Reasoning Ability, Meaningful Learning, and Students' Understanding of Physics Concepts. *Journal of College Science Teaching* 24, 311-14.

Woods, D.R. (1995/96) Teaching and Learning: What Research Can Tell Us. *Journal of College Science Teaching* 25, 229-32.

# Professional Development Standards

Joseph I. Stepans, Maureen Shiflett, Robert E. Yager, and Barbara Woodworth Saigo

**A**s discussed in Chapter 1, the Teaching Standards of the *National Science Education Standards* (NRC 1996) list criteria that can maximize student learning and that are based on research in the cognitive sciences. The Teaching Standards indicate that teaching should be inquiry-based and directed toward specific outcomes; curriculum and assessment should be aligned with the desired "destinations" that students are expected to reach. In this chapter, we consider how university faculty can be enabled through professional development to implement *Standards*-based teaching and why they should try to do so. We also discuss how university science education faculty can design and implement inservice and preservice *Standards*-based professional development programs for K–12 teachers of science.

The Professional Development Standards articulate the criteria for experiences that can lead to significant pedagogical change among teachers of science at all levels. Professional development should focus on how to

- teach essential science content through inquiry,
- integrate knowledge about science with knowledge about learning, pedagogy, and students, and
- develop an ability for lifelong learning.

A quick glance at the *Standards* may suggest that the Professional Development Standards are directed solely to persons who design and implement professional development opportunities for K–12 teachers. However, we believe that the Professional Development Standards are pertinent for every faculty member who is involved in teaching science, for two reasons. First, faculty members are responsible for fostering competence among students majoring in science as well as scientific literacy in science nonmajors. Second, these Standards are based on the assumption that it is the responsibility of teachers of science, kindergarten through college, to engage regularly in professional development during their careers. Professional development opportunities that focus on teaching—whether the participants are K–12 teachers, students in our classrooms, or ourselves—should incorporate understanding of both science content and the teaching-and-learning process.

For many university and college faculty members, a better understanding of the teaching-and-learning process will lead to a change in the way they teach science, enabling them to enhance their students' opportunities to gain both understanding and appreciation of science. Such understanding and appreciation are important for science majors, on whom advances in science will depend. They also are important for nonmajors—the majority of our students—by whom our public policies will be profoundly influenced, from legislation to school board decisions.

Of the nonmajors we teach, there may be no more important group than those preparing to become K–12 teachers. These students will carry the primary message of science understanding to society. It is particularly important that they experience good and varied models of science instruction, rather than just the lectures, verification-type laboratories, and textbook assignments that typify most introductory college science courses.

The Professional Development Standards indicate that successful professional development programs present content information that is grounded in specific instructional and assessment strategies. Professional development experiences that simply talk about other ways to teach, or that focus on demonstrations or specialized laboratory activities, may miss the point that, like our students, we as teachers need concrete, connected experiences to build knowledge, understanding, and ability. Research (e.g., Galloway 2000, Cantrell 2000, Stepans et al. 1999) shows that effective professional development invokes change through direct experiences that put faculty members, anew, in the role of learners taking risks to experience conceptual change. A research-based analysis by Susan Loucks-Horsley and her colleagues (Loucks-Horsley et al. 1987) found that effective professional development, among other things, fosters collegiality, promotes risk taking, and provides reflection time and sustained support.

The most effective professional development for K–12 teachers includes experiences that help them to work with the students in their own classrooms; however, this aspect is commonly neglected when university faculty design and implement professional development experiences for K–12 teachers. Such faculty need to understand the K–12 teaching environment in order to appreciate the challenges and excitement facing teachers at those levels. For example, in a K–12 classroom the role of the teacher differs dramatically from that of the college teacher. There is generally less formal distance between students and teachers and more immediate responses are demanded from both students and teachers—in fact, as we can recall from our own experiences in elementary and high school, teachers and students are engaged with each other nearly every moment. Also, K–12 teachers are expected to appraise and diagnose each child's specific learning aptitudes, needs, and skills, and to provide instruction that is individually appropriate, while dealing at the same time with social and psychological factors. By contrast, in higher education there is a selected population and more of the responsibility for learning is shifted to the student.

In summary, the standards for professional development offer guidance for everyone who designs and provides professional development activities—whether that development is directed to oneself, research scientists, graduate students (among whom future college and university faculty may be found), faculty who teach science courses to undergraduate majors and nonmajors, science educators, or university administrators. In addition, the standards for professional development speak directly to university faculty who provide inservice learning experiences for K–12 teachers.

## Learning Science Content

Professional development for teachers of science requires learning essential science content through the perspectives and methods of inquiry. Science learning experiences for teachers must

- Involve teachers in actively investigating phenomena that can be studied scientifically, interpreting results, and making sense of findings consistent with currently accepted scientific understanding.

- Address issues, events, problems, or topics significant in science and of interest to participants.

- Introduce teachers to scientific literature, media, and technological resources that expand their science knowledge and their ability to access further knowledge.

- Build on the teacher's current science understanding, ability, and attitudes.

- Incorporate ongoing reflection on the process and outcomes of understanding science through inquiry.

- Encourage and support teachers in efforts to collaborate.

*Source:* National Research Council. (1996) *National Science Education Standards.* Washington, DC: National Academy Press, 59.

Professional Development Standard A articulates the criteria that lead to effective learning of science concepts. This Standard—and the *National Science Education Standards* as a whole—calls for a shift in the way K–12 science courses are taught—a shift from emphasis on lectures, textbooks, and worksheets to more emphasis on inquiry and problem solving. Standard A addresses the need for a concomitant shift in K–12 professional development opportunities to include more decision making, active learning, and reasoning: "The vision of science and how it is learned as described in the Standards will be nearly impossible to convey to students in schools if the teachers themselves have never experienced it" (NRC 1996, 56). The science learning experiences for developing teachers that are described in this Standard are all experiences to which university scientists should relate well, because they are the experiences by which they themselves learned and continue to learn their scientific disciplines. The first criterion for learning science content requires that phenomena be studied "scientifically"— a process that involves observations and interpretations of those observations. As instructors and as scientists, we learn best by actively investigating phenomena, asking genuine questions, making precise observations, and interpreting results. The other four criteria for meaningful professional development, which address ways to learn science content, also are a part of the professional life of the scientist:

- *Relevance*: The most successful scientists investigate problems that are timely, potentially useful, and of interest to the scientific and general population. Science learning for prospective and current teachers should be relevant to the kinds of real world connections that are accessible to them and their students.

- *Access to science information*: Knowledge of the accomplishments and failures of prior research is essential for success as a scientist. Reading journals, searching the Internet, and attending scientific meetings are just some of the ways that scientists expand their knowledge, and also are appropriate for teachers at all grade levels.

- *Constructing knowledge*: To be useful to K–16 teachers, science knowledge that comes from their professional development experiences must be connected to their current science understanding. It does little good for teachers—or scientists—to learn interesting but isolated methods and information from their research. The experiences should start at a familiar point, challenge existing ideas and interpretations, and lead to construction of new knowledge and understanding.

- *Collaboration*: It is a rare scientist who does his or her work in isolation. Most successful research breakthroughs have been the collaborative work of a research team. Discussion, debate, review, and critique among peers through a wider science community are all integral to the way scientists, teacher educators, and preservice and inservice teachers add to understanding.

These components of Professional Development Standard A should be incorporated into all professional development efforts that seek to increase the science knowledge of teachers at all levels. Although the need to develop knowledge in science content may be minimal for university faculty, it is probably instructive for all of us to revisit how we learn science. In the following vignette, note that the scientist is put in a position of having to learn the fundamental concepts of a science topic unrelated to his specialty. When this learning occurs alongside a K–12 teacher, both the K–12 teacher and the university faculty member are enriched. The scientist models how science is learned and, in the process, is reminded of the educational power of inquiry.

# From the Field

## Two Perspectives of a Workshop Experience

Maureen Shiflett, Co-Executive Director
The National Faculty

Mr. M. is a fourth-grade teacher in California. Two years ago, he participated in a training program on the use of the "Behavior of Mealworms" module offered by the Pasadena Elementary Science Program of the California Institute of Technology Precollege Science Initiative (CAPSI). CAPSI (*www.nas.edu/rise/example81.htm*), a collaborative between California Institute of Technology and the Pasadena Unified School District (PUSD), provides materials and methods for improving science education. Mr. M. had had minimal education in the sciences as a liberal studies major in college and was insecure about teaching science in his classroom. Consequently, what little science he taught was strictly from the textbook and consisted mostly of having his students memorize terminology with the aid of a few anatomy models. He knew, however, that PUSD was moving toward teaching inquiry-based, hands-on science, and he had heard from other teachers how the training with the science modules had changed their science teaching. At the training session, he had an opportunity to work with the materials in the module in a group of three other teachers and a university scientist. Not only was he able to use the module as his students would, but he enhanced his own scientific knowledge about behavioral science by working with the scientist and the lead teachers who conducted the institute. He was surprised to learn that the university scientist was not a biologist, but a physicist. By working closely with the scientist as they both investigated the behavior of the mealworms, he learned about the scientific processes that scientists use every day. He soon realized that these processes were similar to those his students would use when they worked with the kits provided.

Dr. N. is a physicist at a leading research university. As he went into the workshop, he had had minimal exposure to cognitive research. His knowledge was limited to personal experience in the undergraduate classroom. By working with the teacher as they both investigated the behavior of mealworms, he came to understand the power of inquiry as a learning tool. As Dr. N. continued to work with the Pasadena elementary school teachers, he began to incorporate more inquiry-based teaching into his own university courses. Although many of his class hours were still lecture-based, he identified topics that lent themselves to having students investigate on their own for part of the class time. He also found himself asking more open-ended questions of the students and having them work in cooperative groups.

# Learning How to Teach Science

**STANDARD B**

> Professional development for teachers of science requires integrating knowledge of science, learning, pedagogy, and students; it also requires applying that knowledge to science teaching. Learning experiences for teachers of science must
>
> ● Connect and integrate all pertinent aspects of science and science education.
>
> ● Occur in a variety of places where effective science teaching can be illustrated and modeled, permitting teachers to struggle with real situations and expand their knowledge and skills in appropriate contexts.
>
> ● Address teachers' needs as learners and build on their current knowledge of science content, teaching, and learning.
>
> ● Use inquiry, reflection, interpretation of research, modeling, and guided practice to build understanding and skill in science teaching.
>
> *Source:* National Research Council. 1996. *National Science Education Standards.* Washington, DC: National Academy Press, 62.

The implication of this Standard is that the knowledge of science content alone is not enough for a teacher to be completely effective. Teachers must also use effective teaching strategies. This Standard speaks forcefully to university and college faculty— particularly those who are teaching introductory science courses and courses intended for students preparing to become K–12 teachers. We can conclude from the Standard that courses at that level ought to be taught differently than they typically have been. Humans learn in diverse ways, and diverse approaches to teaching are necessary to maximize learning (Leonard 1997). The most effective learning in science should, wherever possible, have a concrete basis—starting with an observation and a question, followed by establishing and testing hypotheses and analyzing and communicating the results. The Standard also emphasizes that teaching and learning are active, collaborative processes.

The focus of Professional Development Standard B is on both the science content and educational methods courses taught to our developing teachers. It thus requires that those who work with K–12 teachers have deep knowledge of the theories of science learning, pedagogy, and child development. It is a rare university science professor, however, who has such knowledge. Thus a good first step in the self-development of a faculty member's teaching skills—as well as in the planning of an effective professional development program for other faculty at the postsecondary level— is to establish a partnership with the experts in those areas, that is, colleagues in the school of education. Building bridges between those two groups of professionals— those in science disciplines and those in education—is not a trivial task. As a first step, we must recognize and value the expertise both science and education specialists bring to designing effective learning experiences.

The following vignette illustrates how the collaborative process across science disciplines and with education faculty has affected teaching. The major impact has been in courses offered for elementary education majors, but all courses taught by the faculty members involved have been changed in substantive ways.

## From the Field

### A Science Program for Prospective Elementary Teachers
Joseph I. Stepans, Professor of Science and
Mathematics Education, University of Wyoming

A concern for the quantity and quality of science education preparation at the elementary school level brought together faculty members from four different colleges at the University of Wyoming. In Wyoming, as elsewhere, science is frequently neglected or poorly taught at the elementary school level. If they do tackle the subject, teachers often rely heavily on textbooks, worksheets, and recitation, in contrast to the types of learning experiences that have been consistently shown to be more effective: hands-on strategies, cooperative learning, and inquiry/inductive thinking strategies. One contributing factor to this situation is that elementary school teachers frequently have had negative experiences with science themselves and feel (or actually are) underprepared to teach it (Stepans, McClurg, and Beiswenger 1995). To address this problem, faculty members considered ways to improve the preparation for teaching science in the elementary grades.

The faculty members formed teams in areas of subject matter interest: life science, physical science, and Earth science. Each team had members from the College of Education and from other colleges (Agriculture, Arts and Sciences, and Engineering). The leader of each team had some background in both science and education. Most of the faculty members had not worked together before. The teams met together extensively, brainstormed, and eventually developed experimental courses that they felt would be more appropriate for elementary education majors than the traditional offerings. The National Science Foundation provided funding for this planning process.

The experimental courses consisted of *three content courses* (Earth science, physical science, and life science) and *three parallel educational seminars* that students attended concurrently with the respective science courses. A concern of the faculty members was that the content of the courses should not be less rigorous than traditional courses; rather, the content should be presented in alternative ways that would encourage student interest and conceptual understanding of the material. With a maximum of forty students, the content courses were smaller than traditional introductory science courses. They emphasized laboratory activities, group projects, simulations, and field trips. Lecture was de-emphasized.

The purpose of the educational seminars was to connect the concepts the students were studying in the content courses with effective pedagogical methods. During the seminars the students participated in activities appropriate for elementary school children. An evaluation of the program indicated a strong improvement in student attitudes toward science and scientists, confidence in science teaching, and content knowledge (Beiswenger, Stepans, and McClurg 1998). In addition, most students enjoyed the classes. To this day, the mentor teachers

who supervise student teachers in the schools feel that these preservice teachers are better prepared than those in the past, and they appreciate the new ideas and strategies that the new teachers bring to the classroom. The experimental program has been so successful that the university has institutionalized it as a part of its academic program for elementary education majors—both the content classes and the seminar classes are now required.

It is important to note that preservice teachers are not the only beneficiaries of the program. University of Wyoming faculty members appreciate the opportunity to develop and test innovative teaching techniques within the context of the content classes, and these techniques then find their way into all courses that they teach. They report that they also enjoy and benefit from the collaboration with peers in different teaching areas.

Many colleges and universities provide professional development services for their faculty that may be housed in entities called Centers for Teaching Excellence, Teaching and Learning Centers, Programs for Instructional Excellence, or Offices of Teaching Effectiveness. Such centers provide a number of services and resources to help faculty members learn more about teaching and learning and ways to implement diverse strategies. The specific activities differ from center to center, but they generally attempt to provide information, practice, and feedback on ways to improve teaching. Many of the activities are structured to provide a campuswide community for faculty to engage in developing their teaching skills. Faculty development centers offer assistance in most of the following areas:

- Development of presentation materials and use of specific software programs for teaching
- Campuswide seminars and workshops on teaching and learning, featuring both national and local presenters
- E-mail listserv forums and bulletin boards about teaching

- New faculty workshops
- Graduate student teaching workshops
- Teaching circles
- Brown-bag discussions and videotapes
- Development of a professional teaching portfolio
- Peer observation training
- Analysis of instructional effectiveness (materials, methods, assessments)

Faculty development workshops cover teaching strategies that are consistent with the psychology of learning, suggestions on how to lead effective classroom discussions, and assistance in developing new teaching tools, including appropriate uses of technology in teaching. The development of effective classroom assessment techniques to inform instruction and aid in student learning is also emphasized. Most faculty development centers address the scholarship of teaching and ways to increase scholarly growth and productivity, which are important for faculty advancement.

Despite the growing number of faculty professional development centers on college and university campuses, however, significant barriers to systematic professional development for higher education faculty exist. Some of the barriers are related to the *autonomy* of the profession. Each faculty member is in charge of teaching her or his own classes and also may be supervising the teaching of graduate assistants. Another perceived barrier is *time*, in that faculty members and teaching assistants in science courses usually are strongly focused on passing along the content of their own disciplines as thoroughly and efficiently as possible. Some faculty would be receptive to the development of teaching skills, but they are concerned about how they can fit that in along with all of their other responsibilities. *Unfamiliarity* with the abundant research related to cognition and to the teaching and learning of science (as discussed in the Introduction) may be another factor that limits faculty participation in programs aimed at developing teaching skills. *Isolation* is also a factor—professional collegial discussions typically occur at the department or building levels, rather than across campus. Until they experience interacting professionally with colleagues outside their own department, faculty members may not see its value. Recognizing the relative isolation of teaching faculty, most teaching centers create both formal and informal opportunities to bring together faculty and staff from all disciplines, with the improvement of teaching and learning as a focus. In addition, centers take advantage of the power and ease of electronic communication, with e-mail forums and Web-based resources.

The responsibility for overcoming the barriers to faculty development in higher education rests with the development staff as well as with faculty. Audrey Kleinsasser, director of the Center for Teaching Excellence at the University of Wyoming, states that faculty developers need to present theoretically and morally grounded reasons for faculty to change and to improve (Kleinsasser 1999). Moreover, because people are all different, modes of faculty development must be varied (e.g., one-time conferences, long-term faculty study groups, and collaborative efforts involving one or more departments and programs). Staff must create a collegial environment for developing teaching skills and design programs that are relevant and practical.

On the other hand, faculty members bear the responsibility for taking the initiative for professionally developing their own teaching skills. If they acknowledge that teaching is a serious responsibility, they should explore ways to maximize student learning. For teaching faculty, developing teaching skills must become an important part of a total professional development program. In England and Wales, a recent government decision mandates that all new university faculty must complete training programs on teaching and learning (National Committee of Inquiry 1997). According to Gill Nicholls of Surrey University, all new faculty at that institution must experience teaching and learning in different situations (i.e., large-group presentations, individual tutoring, and small-group work). New faculty members are trained in curriculum development, assessment design and implementation, and evaluation of student work. As part of their professional growth, faculty members are encouraged to keep reflective journals and portfolios and are involved in peer observation and feedback.

Whether or not on-campus professional development services are available, increasing numbers of higher education faculty are turning to the Internet and the World Wide Web. A comprehensive, useful, and efficient gateway to diverse activities and different models for the improvement of postsecondary teaching and learning is the Professional and Organizational Development (POD) Network in Higher Education Web site (*www.podnetwork.org*). The network is an active group of individuals who are involved in postsecondary faculty development across the United States. Most of the members are affiliated with university-based offices, although most are not in the sciences. The organization produces a newsletter and an essay series and sponsors an annual conference. As the name suggests, its strength is the active networking of its members, who regularly share ideas and experiences. Many members are available to consult for workshops and faculty development activities at other campuses. The organization's Web site also has links to teaching and learning centers at many colleges and universities in the United States and several other countries. The University of Kansas Center for Teaching Excellence Web site (linked to the POD Web site) includes an excellent list of "Periodicals Related to College Teaching," sorted by discipline, as well as a comprehensive list of "Online University Teaching Centers Across the World." By using links to specific university sites from this page, one can learn a great deal about the philosophy, organization, and activities of the centers.

# Learning to Learn

Professional development for teachers of science requires building understanding and ability for lifelong learning. Professional development activities must

- Provide regular, frequent opportunities for individual and collegial examination and reflection on classroom and institutional practice.

- Provide opportunities for teachers to receive feedback about their teaching and to understand, analyze, and apply that feedback to improve their practice.

- Provide opportunities for teachers to learn and use various tools and techniques for self-reflection, such as peer coaching, portfolios, and journals.

- Support the sharing of teacher expertise by preparing and using mentors, teacher advisers, coaches, lead teachers, and resource teachers to provide professional development opportunities.

- Provide opportunities to know and have access to existing research and experiential knowledge.

- Provide opportunities to learn and use the skills of research to generate new knowledge about science and the teaching and learning of science.

*Source:* National Research Council. (1996) *National Science Education Standards.* Washington, DC: National Academy, 68.

Professional Development Standard C addresses building understanding and developing a climate for lifelong learning by teachers at all levels. It emphasizes the importance of *deliberately examining and thinking about teaching, then using this self-evaluation to improve classroom practice and to continue to grow as a professional.* In other words, college and university science instructors should look upon their teaching as they do their research in science. In so doing, they pursue what Boyer (1990) refers to as "the scholarship of teaching." Investigating ways to use research to improve one's teaching is a valid and powerful scholarly activity for faculty members that also benefits students. The following vignette demonstrates how a professor used a research-based approach to redesign a college biology course.

# From the Field

### Research-Based Change: How One College Professor Approached the Challenge of Changing Teaching

Diane Ebert-May, Professor of Botany and Plant Pathology and Director, Lyman Briggs School, Michigan State University

In the mid-1980s I had an epiphany. I realized that my two children, eight and three years old, were more enthusiastic about learning science than my undergraduates were. Even if I factored the proud-parent effect into the equation, the difference in attitudes about science between my children and my students was striking. So much so that after ten years of teaching courses in undergraduate biology, I was compelled to find a pathway to understand what I thought I understood—science education.

Although I'll spare you the detailed account of events along this path, I want you to know that many mentors influenced my decision to pursue this new interest. Among them was Audrey Champagne, an international leader in science education, and Debbie Smith, a graduate student at the time and now an assistant professor in the Department of Teacher Education at Michigan State University. Both tolerated me (the scientist) and nurtured my understanding and ability to do something that I did not learn in my graduate program in tundra plant ecology—how to study the teaching and learning of science.

In 1989 I moved to Northern Arizona University (NAU) to direct the Science and Mathematics Learning Center and to pursue my interests in biology education. Concurrently, several national reports, *The Liberal Art of Science: Agenda for Action* (AAAS 1989) and *From Analysis to Action* (NRC 1990), added fuel to my fire. In general, these reports advocated the reform of science education at the college level for *all* students, both majors and nonmajors, just as *Science for All Americans* (Rutherford and Ahlgren 1990) had advocated reform of education on the K–12 levels. *The Liberal Art of Science* urged faculty to "teach science as science is practiced at its best" (xi). This statement and subsequent rationale were convincing, so I set about to do just that.

Shortly after my arrival at NAU, I was asked to "fix" the nonmajors biology course, which had an enrollment of approximately seven hundred students per semester. I did not know (although strongly suspected) what was broken, but there was evidence that suggested this "orphan" course was ineffective and disliked by students. Did *all* students in this course have the opportunity to become biologically literate? Why were the lecture halls half full on a given day? Did students have a role in the course? How should I proceed? To me, the answer to the last question was obvious: I would do what I do when I want to pursue a new line of inquiry in plant ecology—write a grant proposal! The proposal I wrote was funded by the National Science Foundation, Division of Undergraduate Education.

The research we conducted compared the biological literacy of nonmajor students enrolled in a traditional introductory biology class (e.g., passive lectures, recipe labs) with that of nonmajors who had enrolled in an experimental class in which they took an active role in their own learning (e.g., cooperative learning in class meetings, inquiry-based laboratories). The results were striking (Ebert-May, Brewer, and Allred 1997; Brewer and Ebert-May 1998; Baldwin, Ebert-May, and Burns 1999). The active learning classroom and laboratory environments became intellectually stimulating and challenging, and the majority of students learned.

What were the key components of this successful innovation? I followed the pathway I know as a scientist, and learned from the literature why studying humans is much more unpredictable than studying plants. Cognitive science and human behavior guided my reading and thinking. Research questions that used both quantitative and qualitative research designs helped me gather the appropriate evidence that would convince my peers and me that my students had achieved the goals of the course.

I chose not to merely tinker with my course; rather, I "trashed" any component that promoted passive learning.

Moreover, I "trashed with confidence," based on the literature about cooperative learning, cognitive development, and constructivism (a view of how people construct knowledge). Can tinkering with one intervention at a time improve instruction? It depends on the intervention. If the change in instructional design were to address the core issue of active learning by *all* students, then I predicted student gains in learning. If the instructional change was not substantive or was short-lived, I predicted no difference in student learning.

I studied assessment—that is, data collection with a purpose. I aligned the learning goals of what I wanted my students to know and be able to do with the types of assessments I used to gather data. Thus, to demonstrate their abilities to be problem solvers about biological issues, my students solved problems. My students' written and oral explanations gave me both insight about their understanding and substantive feedback to improve my course.

Importantly, I did not embark on this adventure alone. My faculty and graduate student colleagues over the years from around the country formed the "cooperative group" that together took risks, tested strategies, listened to students, cheered about successes, and agonized over failures. We applied the same level of significance to the conclusions we made about teaching as we did in our research. We supported our passions for teaching and learning and had the courage to do something about it.

University faculty who are involved in research in any field, whether in a science discipline or in education, continually hone the habits of mind and skills that stimulate and sustain their learning. Faculty at two-year institutions or those for whom teaching is virtually their sole professional responsibility must find time and energy (in addition to that required to meet teaching obligations) to read journals, attend professional meetings, and otherwise remain active in a research area. The bulleted items in Professional Development Standard C remind us that learning requires access to information, practice, feedback, analysis of data, and reflection. These tasks are facilitated by sharing expertise among peers and the use of role models, but they require adequate time to carry out. Much of Professional Development Standard C falls in the purview of the administration, in that it is the administration's responsibility to provide adequate opportunities for professional development.

At colleges and universities, administrators provide for professional development in different ways. Most opportunities for science faculty involve engaging in new research experiences in the discipline. However, the opportunities to engage in experiences designed to improve teaching are increasing and becoming more visible. The use of faculty mentors, the growth of teaching-and-learning centers, and feedback from colleagues are becoming more common. If an institution does not provide such support, faculty are looking to professional societies for guidance on ways to improve teaching undergraduate students (Siebert et al. 1997).

In K–12 schools, most intensive professional development activities occur during the summer. University faculty members can participate in such activities. One valuable experience that scientists can offer K–12 science teachers is the opportunity to engage in scientific research. Although the *National Science Education Standards* emphasize teaching science by inquiry, most K–12 teachers have never had the opportunity to do scientific research. Consequently, they lack confidence (Fraser-Abder and Leonhardt 1996) and are unprepared to guide their students in the types of critical thinking that scientific inquiry requires.

Melear et al. (1998) suggest that research experiences can "enculturate teachers into science and…foster scientific behavior and thinking." They describe a course in which preservice secondary science teachers participated in a semester-long research experience facilitated by university scientists and science educators. At the outset of the course, the preservice teachers were provided with an "unknown" organism, *C-Fern* (*Ceratopteris*) and were told to find out something about it. They experienced considerable tension during the initial meetings of the course. Although all of the preservice teachers were senior or post-degree biology majors, they were unfamiliar with the processes of science and unused to the lack of structure demanded by the course. The need to work together efficiently was also a source of tension. The authors remarked that "becoming a scientist, or going from a dependent learner to a self-directed and independent learner, is inherently fraught with frustrations, ambiguities, and confusion." The preservice teachers eventually overcame the initial barriers and produced significant work, including a study accepted for publication in

*Bioscience.* By the end of the course, they showed development in scientific thinking and gained confidence in their ability to do science; all had a positive attitude toward the experience and felt that it would have a positive impact on their teaching. One of the chief complaints of these preservice teachers was that they had studied science for so long without having had the opportunity to do the kind of research required by the course.

The *Standards* view professional development as a continuous, lifelong process that emphasizes decision-making, theory, and reasoning. These are process skills often taught in introductory science courses at the university level and should, therefore, provide the basis for lifelong learning of science. A focus on these problem-solving skills will serve both our future teachers and our science majors. The following vignette illustrates how faculty at the University of Wisconsin have determined how successfully a new introductory chemistry course has addressed the learning of these critical lifelong skills.

## From the Field

### Analysis of an Innovative College Chemistry Course

Barbara Woodworth Saigo, President, Saiwood Biology Resources
Susan Millar, LEAD Center Director, University of Wisconsin–Madison

The University of Wisconsin (UW)–Madison's Learning through Evaluation, Adaptation, and Dissemination (LEAD) Center helps faculty and staff members who seek ways to foster effective learning by:

- Conducting evaluation research on student learning environments, processes, and outcomes.
- Helping faculty and staff understand the organizational and cultural issues that are involved in implementing new approaches.
- Disseminating evaluation and adaptation research findings locally and nationally.

The UW–Madison established the LEAD Center in fall 1994 to provide third-party evaluation research in sup-port of educational improvement efforts at both undergraduate and graduate levels. Center funding comes from internal and external grants obtained by its clients or in partnership with clients. Thus far, LEAD has completed more than forty projects. Almost all center clients have been science, engineering, and mathematics faculty who are exploring ways to improve education by using strategies that foster more active student learning.

The LEAD Center approach to evaluation is client-driven and student-focused. Both qualitative and quantitative research methods are employed, often involving multiple methods, sources of information, and researchers. LEAD evaluation documents that clients wish to make widely available can be found at *www.engr.wisc.edu/~lead*.

Examples of how the LEAD Center's evaluation research helps science faculty improve student learning can be found in articles and reports on the New Traditions Chemistry Systemic Reform project at UW–Madison, a five-year, NSF-funded project that began in 1995. Of the articles, those focusing on a freshman analytical chemistry course, Chemistry 110, are of particular interest. The faculty member who designed a new version of this course wanted students to develop a deep appreciation of scientific problem-solving processes and critical thinking skills. To accomplish this goal, he used a broad range of active, group-learning strategies.

The LEAD evaluation of the New Traditions Chemistry 110 course consists of a detailed case study. The study is placed into a more meaningful context by a comparison case study of a different section of the course taught by a faculty member who had received teaching awards and who used the more conventional lecture-lab-discussion format. Oral assessments of the students in these two sections of the course conducted by twenty-five faculty colleagues from across the university showed that the New Traditions faculty member had achieved his goals in terms of faculty colleague judgments (Wright et al. 1998). Specifically, the faculty assessors ranked the students

from the New Traditions section more highly in terms of specific learning attributes. The differences between the two sections of the course were statistically significant.

A second study of this course was based on surveys completed by three cohorts of Chemistry 110 students in their junior and senior years who were asked about the impact of the course on their college experiences. Among the findings were the following:

- Two and three years after the course, students from the New Traditions section gave the course high marks on outcomes sought by their instructor.
- They reported notably high levels of interest, enthusiasm, and confidence in their learning.
- There was strong positive correlation between students' affective responses and their assessment of the cognitive impact of Chemistry 110 on their learning process and on their skill attainment.
- Women reported especially strong benefits from the New Traditions teaching methods.

These evaluation findings are valuable to the New Traditions faculty and their college and university colleagues around the nation because they provide (1) strong evidence that constructivist teaching methods can have the desired effects on student learning outcomes, and (2) detailed insights into how and why these outcomes are achieved.

# Planning Professional Development Programs

Professional development programs for teachers of science must be coherent and integrated. Quality preservice and inservice programs are characterized by

- Clear, shared goals based on a vision of science learning, teaching, and teacher development congruent with the *National Science Education Standards*.

- Integration and coordination of the program components so that understanding and ability can be built over time, reinforced continuously, and practiced in a variety of situations.

- Options that recognize the developmental nature of teacher professional growth and individual and group interests as well as the needs of teachers who have varying degrees of experience, professional expertise, and proficiency.

- Collaboration among the people involved in programs, including teachers, teacher educators, teacher unions, scientists, administrators, policy makers, members of professional and scientific organizations, parents, and business people, with clear respect for the perspectives and expertise of each.

- Recognition of the history, culture, and organization of the school environment.

- Continuous program assessment that captures the perspectives of all those involved, uses a variety of strategies, focuses on the process and effects of the program, and feeds directly into program improvement and evaluation.

*Source:* National Research Council. (1966) *National Science Education Standards.* Washington, DC: National Academy Press, 70.

The remaining segment of this chapter provides selected examples of professional development programs, especially those that involve collaboration of university and K–12 faculty. The examples correspond to Professional Development Standard D, which specifically recommends *coherence*, *integration*, *coordination*, *collaboration*, and *continuity*. The examples demonstrate that the approaches called for in the *Standards* are both feasible and successful.

Before looking at the examples, it is important to note that the research on which the *Standards* are based emphasizes (as discussed in the Introduction) a constructivist view of learning—that is, that true learning results from conceptual change and that this philosophy should be at the heart of any high-quality professional development process. As represented by Piaget (1971), a learner is initially comfortable (at equi-

**Figure 2.1**

*Wyoming TRIAD Professional Development Process: Conceptual Change through Personal Experiences*

| Initial equilibrium | Disequilibrium | New equilibrium |
|---|---|---|
| Teachers have preconceptions about what students know and how they learn. They are comfortable with present ideas about what should be taught and how, and with traditional curriculum, instructional strategies, and assessments | Student interviews reveal discrepancies between student understanding and teacher assumptions. Teachers use what they learn as a basis to implement new instructional and assessment strategies and do classroom research. | Through modeling and personal experience, teachers find more effective ways to stimulate and assess learning. They become confident of a professional knowledge base, its relevance and applicability. They use research in curricular and instructional decisions. |

*Note:* For a description of the Wyoming TRIAD Professional Development Process, see pages 46–50.

librium) with an existing concept. When new experiences and information conflict with that existing belief, mental conflict and disequilibrium occur. Eventually, through repeated experiences in which the learner's prior belief is challenged, the learner may construct an altered understanding, reaching a new equilibrium with respect to the idea. Piaget refers to the movement from equilibrium through disequilibrium to new equilibrium as *equilibration*. Application of this idea to a professional development program fosters conceptual change by providing new information, greater understanding, and new skills and attitudes through firsthand, concrete experiences. Figure 2.1 shows how conceptual change toward students, teaching, and learning occurs through the diverse approaches experienced in one *Standards*-based professional development process.

Some of the most effective professional development models are collaborative partnerships between K–12 and university faculty and administrators. Professional development that is consistent with the vision of the *Standards* requires that university faculty working with K–12 teachers rethink the relationship, acknowledging that both groups can learn from each other. Collaboration can simultaneously improve and update the understanding of science by K–12 teachers and help postsecondary faculty to understand how scientific information must be translated into developmentally appropriate curriculum and learner-centered pedagogies.

Working with elementary and secondary teachers also can help college and university faculty to better understand the nature of learning and the development of learners. A fundamental rationale is that faculty whose primary responsibility is teaching, at any level, should be scholars of both the content area and of the teaching process. The merging of content knowledge and teaching knowledge is referred to as *pedagogical content knowledge* (Shulman 1986). It refers to the methods a teacher uses to make the content more understandable to the learner—for example, the use of analogies, demonstrations, and examples. This emphasis insists that well-designed, effective educational experiences for students be based on the findings of cognitive research and developmental psychology. To do less is to ignore scientific advances in these fields. An analogy might be teaching biology based on what was known thirty years ago rather than on what is known today.

An important concept in *Standards*-based professional development is the *community of learners*. When university faculty who have common goals reflect on and share ideas among themselves over an extended period of time, the growth can be an actual shift in the departmental culture. Old and new ideas and practices are dissected and evaluated through questioning, testing, and consideration of multiple ideas and points of view. There is time to engage in experiences to help faculty become "constructivists of the teaching-and-learning process" (Stepans, Saigo, and Ebert 1999). An important side effect of programs that seek to change the culture of graduate education, and thus undergraduate education, may be to encourage university faculty to interact professionally with colleagues to improve teaching.

Four examples of professional development programs that follow guidelines given in the Professional Development Standards are described below.

## The Iowa Chautauqua Model

The Iowa Chautauqua model (Blunk and Yager 1990; Yager and Blunk 1991; Yager 1993) is a staff development model that meets the criteria described in the *Standards*. The model involves a yearlong process that includes the four major parts outlined in Figure 2.2. The model requires participants (mostly K–12 teachers, but also curriculum developers, administrators, and even some parents) to address questions about learning and establishes a meaningful learning context by involving students in asking questions, providing answers to proposals, designing experiments, amassing evidence, and communicating results. The Iowa Chautauqua model is a variation of the well-tested professional development Chautauqua Short Courses funded by the National Science Foundation (NSF 1999). Those courses employ college science teachers as resource people who work with K–12 teachers. College teachers first go through training in pedagogy prior to their participation in the project. Subsequently, while working closely with K–12 teachers, they learn methods that are applicable to their own teaching. The courses, which aim to enable undergraduate teachers in the sciences to remain current in their teaching, generally meet at university and college sites with approximately twenty-five college teachers for three successive days of full-time activity; a few courses include a one- or two-day follow-up after an interval of roughly six months. Faculty members are selected to participate only with the encouragement of their home institutions, and the courses provide knowledge and techniques that can be immediately applied to their teaching.

## Figure

*The Iowa Chautauqua Staff Development Model*

### Part One: CHAUTAUQUA LEADERSHIP CONFERENCE

Lead teachers (i.e., K–12 teachers with experience in the methods and materials of the conference) meet to
- plan summer and academic year workshops,
- gain greater familiarity with target curricula,
- enhance instructional strategies and leadership skills, and
- refine assessment strategies.

### Part Two: THREE-WEEK SUMMER WORKSHOPS

Lead teachers, university faculty, and local scientists work with K–12 teachers at on-site workshop settings. Teachers are introduced to constructivist teaching as depicted in the six NSF Target Curricula. Teachers:
- Participate in activities and field experiences that integrate science concepts and principles found in the target programs
- Make connections between science, technology, and society in the context of human experiences exemplified in the target curricula
- Use local questions/problems/issues such as air quality, water quality, and land use and management as the context for conceptual development
- Select specific modules and learning activities for the initial pilot work in September

### Part Three: FIRST CLASSROOM TEACHING TRIAL

K–12 teachers involved in summer workshops teach and assess a pilot unit from the target program using constructivist practices.

### Part Four: ACADEMIC YEAR WORKSHOP SERIES

Lead teachers, university faculty, and local scientists work with K–12 summer teachers and new teachers.

***Fall Short Course. Twenty-hour instructional block. Activities include:***
- Reviewing problems with traditional view of science and science teaching
- Outlining the essence of constructivist teaching and learning
- Defining techniques for selecting teaching modules and assessing their effectiveness

- Selecting a tentative sequence of topics
- Practicing specific assessment tools in multiple domains
- Analyzing current practices in relation to constructivist practices
- Establishing a central supply system to support implementing target curriculum

### Interim Projects. Focusing on three or four units from target curriculum project(s).

- Collection of pre-instruction data from students
- Identification of needed input from developers and local staff
- Interaction with teachers from same grade level
- Integration and articulation of problems
- Review of materials on similar topics from other targeted programs

### Spring Short Course. Twenty-hour instructional block. Activities include:

- Analyzing constructivist experiences in grade-level groups
- Discussing assessment results focusing on student learning
- Reflecting and analyzing teaching changes related to constructivist practices and use of new materials
- Considering new information about constructivist practices and additional materials
- Planning next steps for expanding practice for use of even more modules from target curricula
- Planning for involvement in professional meetings and local school transformations
- Structuring a complete framework for given grades and across grades

## The National Faculty

When university faculty design and implement programs for K–12 teachers, K–12 teachers and administrators must be involved as key players in the planning stages. Successful examples of this type of planning are The National Faculty (TNF) programs (*www.tnf.org*) for developing K–12 teachers. These programs link teachers with leading college and university scholars in many disciplines, focusing on content and discipline-based inquiry. The first step in creating a program is a meeting between the TNF staff and representatives from a school district about the district's professional development needs. Next, TNF staff bring together experienced university scholars, lead teachers, and district curriculum experts to map out a strategy, which usually includes two or three intensive summer institutes with follow-up activities during the academic year.

After this collaboration stage, appropriate university scholars are recruited. Then, a step occurs that has often been overlooked in professional development programs using university faculty: Extensive professional development on the needs and mi-lieu of K–12 teachers is provided *for the university scholars* before they begin to work with teachers. For example, TNF has begun an initiative in conjunction with California State University, Long Beach, and Long Beach Unified School District (LBUSD) to orient ten newly recruited mathematics scholars about how they can best assist their K–12 colleagues in meeting their professional development goals. Mathematics professors have observed experienced scholars conduct workshops with teachers; they have visited a variety of high school mathematics classrooms in the company of the district mathematics coordinator; and they have attended seminars and discussions about national, state, and district standards for mathematics education. The culminating experience will be a summer workshop for selected LBUSD teachers, designed and implemented by the mathematics scholars. At the end of the year-long program, funding will be sought to have these ten mathematicians provide ongoing professional development opportunities for LBUSD mathematics teachers.

## The Wyoming TRIAD

The Wyoming TRIAD (WyTRIAD) is a teacher inservice experience that explicitly demonstrates the Professional Development Standards (Figure 2.3). It is based on an ongoing, three-way partnership among the teachers at a school, their principal (building administrator), and a professional development facilitator (usually a university faculty member). In the WyTRIAD, participants collaboratively experience—not just read or hear about—an integrated set of research-based components. Through modeling, practice, reinforcement, and follow-up, they experience conceptual change, constructing new knowledge, developing new skills, and revising their philosophy of education (Stepans, Saigo, and Ebert 1999).

As will be described in the following vignette, WyTRIAD is school-based, job-embedded, contextual, and collaborative and takes place over an extended period of

## Figure 2.3

*Incorporation of the* National Science Education Standards *into a Professional Development Model*

| National Science Education Standards call for... | WyTRIAD... |
|---|---|
| • clear, shared goals | • is based on mutually developed goals |
| • integrated components | • consists of fully integrated components |
| • coordinated components | • includes a coordinated plan and schedule |
| • developmental approach | • uses a constructivist approach |
| • individual and group-specific elements | • is nonprescriptive, based on teachers and school |
| • collaborative basis | • is an interactive, collaborative partnership |
| • relevance to school site | • occurs on-site and is classroom-based |
| • continuous assessment | • involves ongoing action research and evaluation |

time. It emphasizes collegiality and classroom-based action research. Its fundamental strength is the committed, active participation of all three partners. The model is flexible enough to fit in as the functional delivery mechanism for a variety of comprehensive, systemically oriented professional development plans.

The WyTRIAD process consists of four or five sessions and assigned activities to be conducted in the classroom between the sessions. One cycle takes place over an entire semester (four to five months). Each session is one or two days in length and occurs at the school, with classroom modeling. During the between-the-session intervals (two to six weeks), teachers implement and evaluate the impact of targeted strategies for instruction, curriculum evaluation and development, and peer collaboration. In this way, changes in teaching are based on classroom research, with immediate feedback. Additional aspects of this model are provided in the vignette that follows.

## From the Field

### The Wyoming TRIAD (WyTRIAD) Professional Development Process

Barbara Woodworth Saigo, President, Saiwood Biology Resources
Joseph I. Stepans, Professor of Science and Mathematics
Education, University of Wyoming

Several goals support the WyTRIAD's vision of transforming classrooms. The professional development model provides opportunities for K–12 teachers and administrators to

- Become aware of the research on teaching and learning.
- Conduct classroom research and apply it in their own classrooms.
- Become aware of specific science and mathematics misconceptions that they and their colleagues may hold.
- Identify student preconceptions and misconceptions as a basis for designing appropriate expectations, experiences, and assessments.

- Make decisions based on their own classroom research about what concepts are appropriate at various levels.
- Learn about, see modeled, implement, and evaluate classroom strategies, ideas, and activities.
- Participate in peer sharing, coaching, and collaboration.
- Become more reflective in regard to both classroom experiences and their own personal and professional growth.
- Develop strong collaborative teams in schools, with teachers, administrators, and professional development partners all working together over an extended period of time to

improve the learning experiences of children.

Activities chosen to meet these goals lead participants through critical stages of awareness, construction and implementation of knowledge, and evaluation of self, the classroom, and the curriculum. Significantly, *the principal attends all sessions with the teachers* so he or she will understand fully what the teachers are doing and provide needed support for the teachers to collaborate and implement their assignments between sessions.

The constructivist view that true learning results from conceptual change is at the heart of the WyTRIAD. Over the months, participants test their existing ideas through experiences with the children in their own classes; they go through the cognitive process described as *equilibration* as their existing ideas are challenged (Figure 2.1). When teachers and administrators come into the program, they are comfortable—at equilibrium—with their beliefs about subject-matter content, teaching, assessment, what students know at a given level, and how they learn. During the sessions, they learn about research, explain and question their views and practices, participate in lessons, and observe lessons modeled by the facilitator with their own students. They analyze student responses to the modeled lessons. Teachers also interview students about a topic they are preparing to teach and record student responses. Often the students' preconceptions are surprisingly at odds with the teachers' assumptions about what they know. As a result,

the teachers begin to question their views; they then use what they have learned to design lessons using a conceptual change strategy. Before the next class, they teach the lesson and record observations of student responses.

At each meeting, participants share their observations and borrow and learn from each other and the facilitator. Throughout the process they document their experiences, observations, reflections, and ideas. Midway through the process, teachers are challenged to apply their new understandings to other content areas, perhaps by designing integrated lessons or examining the district curriculum. They then are expected to raise new questions and to continue their learning. By the end of the process, they are approaching a new equilibrium, consisting of what they now understand about the teaching-and-learning process and believe to be appropriate curriculum, instruction, and assessment.

The *National Science Education Standards* view teaching as collegial and intellectual; they support a vision of restructured schools and schools of education in which "prospective, new, and veteran teachers are conducting school-based inquiry, evaluating programs, and studying their own practices—with one another and with university-based colleagues" (Darling-Hammond 1996). In this and other ways, the WyTRIAD doesn't just teach about national standards, it embodies them. Teachers are actively involved in research on teaching effectiveness and use their findings from the research to inform and modify their teaching. Teacher research in WyTRIAD includes interviewing students to identify preconceptions and misconceptions, implementing and tracking the effectiveness of new strategies, observing students and recording observations, keeping track of changes in students and in themselves, examining assumptions of curriculum and textbooks, sharing and borrowing ideas from colleagues, and engaging in peer collaboration and coaching.

Does the WyTRIAD work? Studies of its impact on teachers and schools (Stepans and Saigo 1993; Kleinsasser and Miller 1994; Cantrell 2000; Galloway 2000) concluded that the WyTRIAD is a successful model that challenges teachers to rethink and revise their approach to the teaching-and-learning process, increases their confidence in teaching science and mathematics and in making curricular decisions, and effectively introduces them to constructivism in both teaching and learning. Interviewing students enlightens instruction and improves communication with students. The conceptual-change teaching strategy is regarded as highly successful. Peer coaching and sharing create supportive relationships that are highly valued, as is the relationship with the university person. The studies also found that the key element to success within a school is the active participation and support of the building administrator. A collaborative study of high school biology students demonstrated a significant difference in retention of concept knowledge by students

in sections that employed the philosophy and strategies emphasized in the WyTRIAD (Saigo 1999).

Effectiveness also may be inferred from research on other models that incorporate some of the same features, including site-based management (Wohlstetter 1995) and site-based professional development (Sykes 1996; Joyce and Shower 1996). In regard to teacher collaboration, Anderson (1995) reports that "nothing we saw in our case studies showed more influence for productive change than [daily] collaboration among teachers." Swanson (1995) affirms the concept of teacher development through partnerships that seek "continuous improvement based on reflection, evaluation, and ongoing research." The role of the building administrator as a key partner in fostering a collaborative, open school community, in which there is trust among teachers and between teachers and administrators is emphasized by Hoerr (1996).

When the WyTRIAD teachers observe modeled practices, design ways to implement them, and gather data on what they are learning from their students and from each other, the teachers are in what Ball (1996) calls "a stance of critique and inquiry toward practice." It contrasts to the stance "that concentrates on answers: conveying information, providing ideas, training in skills," which is the model for traditional inservice workshops, publications, and professional meeting sessions. The inquiry stance involves examining and debating new ideas with others, "considering how other resources and knowledge might be useful," and adapting and generating new knowledge. Gough (1996) also remarks on the importance of providing teachers with opportunities to construct their own knowledge in order to become true agents of reform.

In addition, the WyTRIAD process matches advice cited by Anderson (1995) that reformers think systemically, focus on matters of student learning, make teacher collaboration the foundation of their work, and provide the support that teachers need.

## Preparing Future Faculty

Corcoran (1995) has synthesized a list of guiding principles for designing professional development opportunities for K–12 teachers that are also relevant for university faculty members who wish to develop their own teaching skills. Foremost among these principles is that successful professional development activities are linked with the institution and policies already in place. They must be planned in equal partnership with the faculty and administrators who will be involved. Such planning allows for shared goals to be articulated and maximizes the possibility that program components will be integrated into the faculty members' teaching.

An example of this approach, where development is sustained over time and includes a variety of experiences, occurs in the Preparing Future Faculty (PFF) programs that are sponsored by the Association of American Colleges and Universities (AAC&U) and the Council of Graduate Schools and funded by a grant from The Pew Charitable Trusts (for program description, see *www.preparing-faculty.org*). There are currently PFF programs in roughly seventy-five research institutions and in a variety of disciplines; all have the purpose of improving the way future faculty are prepared for teaching, research, and service responsibilities. Each research institution works with a cluster of undergraduate institutions to design and pilot models for integrating carefully planned preparation for faculty careers into existing graduate programs (AAC&U 1998). There are over three hundred institutions involved in the various clusters; these institutions expose PFF graduate students to a wide variety of faculty environments, including those with a strong liberal arts focus, comprehensive universities, historically black and Chicano universities, and two-year colleges.

In 1998, twenty-one of the PFF programs were established through funding from the National Science Foundation's initiative Shaping Future Faculty in Mathematics and the Sciences. Collaborating with professional societies, these PFF programs focus on graduate student development in departments of chemistry, physics, computer science, and mathematics at research institutions. The following vignette describes the planning and development of a PFF program at Duquesne University. The Duquesne program began with students in the 1999–2000 academic year. Although there is ongoing assessment of activities, the full effect of the program will not be known for two or three years. The vignette illustrates the process of planning a professional development program at the university level.

# From the Field

## Preparing Future Faculty

David Seybert, Department of Chemistry
Duquesne University

The Ph.D. is the accepted terminal degree for college/university science faculty. It is, however, a research degree rather than a multidimensional preparation for careers in academia, even though teaching is the primary assignment for most faculty members. Ph.D. students receive little exposure to pedagogical issues or curriculum development and innovation; even duties as a graduate teaching assistant typically involve teaching pre-defined lab curricula. For many Ph.D. chemists entering academia, the first years as an assistant professor may be spent learning how to be a faculty member through trial and error.

The sciences in general, and chemistry in particular, must take adequate steps to ensure that 21st century college and university students are educated by the most effectively prepared faculty we can produce. Toward this goal, we are integrating into our chemistry and biochemistry Ph.D. program the necessary infrastructure to produce high-quality faculty members for chemistry education. Our program will become part of the Shaping the Preparation of Future Science and Mathematics Faculty initiative, a Preparing Future Faculty (PFF) program of the Council of Graduate Schools (CGS) and the Association of American Colleges and Universities (AAC&U).

In spring 1999, the CGS and AAC&U, with support from the National Science Foundation, invited professional scientific societies to join in effecting change in graduate programs to prepare students for careers in the academy. The American Chemical Society selected Duquesne University (DU) as one of five research institutions to participate in PFF programs in chemistry, and DU designed a PFF program that could be incorporated into a newly revised Ph.D. program.

Our revised Ph.D. program in chemistry and biochemistry enrolls first-year students in several short, intensive skill courses followed by or concurrent with two semester-long research rotations. Students matriculate to Ph.D. candidacy at the end of their first year, upon recommendation of their committee. We created additional flexibility by eliminating a required minimum number of credits. The average student accumulates forty to fifty credits, but now has opportunities to include novel educational opportunities, such as the Preparing Future Faculty program. Mentoring is the key component of our program.

In designing our program, we studied our recent Ph.D. graduates who have established successful academic careers, identifying key opportunities and experiences that appeared to be important factors in their success. Many of these and related activities became part of our design. We also enlisted neighboring institutions to become partners in training and mentoring aspiring faculty. Western Pennsylvania is rich in high-quality, two- and four-year institutions with faculty having background and expertise in diverse undergraduate chemical education environments. Our initial partners were from Chatham College (Pittsburgh), Community College of Allegheny County (Pittsburgh), La Roche College (Pittsburgh), Seton Hill College (Greensburg), St. Vincent's College (Latrobe), and Thiel College (Greenville). The PFF program is administered by a steering committee of partner institution faculty members and second- or third-year Ph.D. candidates, selected by the steering committee on a competitive basis.

Although our traditional doctoral program is strong in preparing graduate students to carry out research and make professional presentations, attention to teaching and college service are minimal. Our PFF program, on the other hand, addresses teaching, research, service, and professional activity. Activities designed to strengthen these important expectations of future faculty include:

- *Monthly brown-bag lunch discussion groups*. Mentors, other faculty, and students discuss educational and pedagogical issues. Faculty mem-

bers lead initial meetings and focus on their own experiences, such as career paths, unique opportunities and problems they have encountered, and creative solutions and activities they have developed. Later during the fall semester, student participants present and lead discussions on such topics as distance learning in chemistry, the role of undergraduate research in chemical education, teaching chemistry to nonscience majors, and using problem-based learning and cooperative learning.

- *Use of Center for Teaching Excellence facilities*. Duquesne University's Center for Teaching Excellence organizes weeklong training and orientation sessions in August for all new teaching assistants. PFF students also have full access to more than five hundred books, articles, and videotapes on college teaching; forty-five to fifty faculty workshops; and other appropriate CTE activities.

- *Faculty mentors*. Each student is paired with a faculty mentor from a partner institution and works closely with the mentor to explore teaching, research, and service roles at the institution. Students share their diverse experiences during the brown-bag lunches. The faculty mentors are appointed as Visiting Scholars at DU, with access to library facilities, research instrumentation, seminars, lectures, and colloquia.

- *Student projects.* During the spring semester, each student develops a chemistry education module, involving new or modified experiments to be incorporated into existing undergraduate laboratory courses, a molecular modeling activity, a distance learning module, or some combination of these. Students present their projects during our annual capstone colloquium.

- *Annual colloquium.* Each summer, students present the results of their projects. A nationally or internationally renowned chemical educator presents the plenary lecture. In addition, recent, successful Ph.D. alumni relate how their graduate education helped them to build successful academic careers. Faculty mentors are encouraged to invite undergraduate students to the colloquium.

The PFF program has involved all member institutions and graduate students throughout the planning phase. Important principles from the *National Science Education Standards* that have been incorporated into planning our program include the fostering of a community of learners, exposure to research findings in learning, time to practice and evaluate alternative methods of teaching, and exposure to diverse learning environments and student populations.

If you are an individual searching for new and more effective ways of teaching, how do you plan professional development experiences that will allow you to change in a way that is consistent with the vision of the *National Science Education Standards*? Some guidelines that appear throughout the chapter are summarized here.

**1** *Familiarize yourself with the types of experiences that lead to effective learning.* Based on research in the cognitive sciences, the *Standards* cite active, inquiry-based, and collaborative experiences as important modes of learning.

**2** *Explore teaching strategies consistent with promoting these modes of learning.* Design learning activities that require active student participation, laboratories that lead students to "discovery," and group projects. It's okay to begin with small steps by incorporating a few of these activities into existing courses.

**3** *Collaborate with others.* Work with other members of your department; if you are the only person in a department or program interested in developing teaching skills, connect with teaching faculty in other departments—the department of education at your institution might be a good place to start. Or connect with teaching faculty from other institutions through participation in professional teaching societies (e.g., the Society for College Science Teachers, the National Science Teachers Association, and education divisions of discipline-specific societies).

**4** *Reflect on your teaching.* Measure the effect on student learning of what you do, and use those assessments to guide your teaching. Many teaching strategies are not successful on the first try and must be modified over time.

**5** *Persevere*. Learning to teach more effectively is a developmental process, and your efforts must be sustained over time. For inspiration, read about the frustrations and solutions to implementing change that have been documented by others involved in reform (Caprio 1997).

**6** *Plan your professional development activities*. Professional development experiences that are coherent and relevant to your teaching responsibilities are more likely to lead to lasting change.

## References

American Association for the Advancement of Science (AAAS). (1990) *The Liberal Art of Science: Agenda for Action*. Washington, DC: AAAS.

Association of American Colleges and University (AAC&U). (1998) *The Impact: Assessing Experiences of Participants in the Preparing Future Faculty Program 1994-96*. Pruitt-Logan, A. S., Gaff, J. G., and Weibl, R. A. A series of occasional papers. Washington, DC: AAC&U.

Anderson, R. D. (1995) Curriculum Reform: Dilemmas and Promise. *Phi Delta Kappan* 77(1), 33-36.

Baldwin, J., Ebert-May, D., and Burns, D. (1999) The Development of a College Biology Self-Efficacy Instrument for Non-Majors. *Science Education* 83(4), 397-408.

Ball, D. L. (1996) Teacher Learning and the Mathematics Reforms: What We Think We Know and What We Need to Know. *Phi Delta Kappa*n 77(9), 500-508.

Beiswenger, R. E., Stepans, J. I., and McClurg, P. A. (1998) Developing Science Courses for Prospective Elementary Teachers. *Journal of College Science Teaching*, 253-57.

Blunk, C. K., and Yager, R. E. (1990) The Iowa Chautauqua Program: A Model for Improving Science in the Elementary School. *Journal of Elementary Science Education* 2(2), 3-9.

Boyer, E. L. (1990) *Scholarship Reconsidered: Priorities of the Professorate*. Princeton, NJ: Carnegie Foundation for the Advancement of Teaching.

Brewer, C. A., and Ebert-May, D. (1998). Hearing the Case for Genetic Engineering: Breaking Down the Barriers of Anonymity through Student Hearings in the Large Lecture Hall. *Journal of College Science Teaching* 28(2), 97-101.

Cantrell, P. (2000) The Effects of Selected Components of the WyTRIAD Professional Development Model on Teacher Efficacy. Doctoral diss., University of Wyoming.

Caprio, M. (Ed.) (1997) *From Traditional Approaches toward Innovation*. The SCST Monograph Series. Tigerville, SC: Society for College Science Teachers.

Corcoran, T. (1995) *Helping Teachers Teach Well: Transforming Professional Development*. CPRE Policy Briefs. Consortium of Policy Research in Education. New Brunswick, NJ: Rutgers University.

Darling-Hammond, L. (1996) The Quiet Revolution: Rethinking Teacher Development. *Educational Leadership* 53(6), 4-10.

Ebert-May, D., Brewer, C. A., and Allred, S. (1997) Innovation in Large Lectures—Teaching for Active Learning through Inquiry. *BioScience* 47(9), 601-607.

Fraser-Abder, P., and Leonhardt, N. (1996) Research experiences for teachers. *The Science Teacher* 63(1), 30-33.

Galloway, D. (2000). The Impact of WyTRIAD Professional Development on Teacher Change. Doctoral diss., University of Wyoming.

Gough, P. B. (1996) No More "Pie in the Sky." Editorial. *Phi Delta Kappan* 77(7), 459.

Hoerr, T. R. (1996) Collegiality: A New Way to Define Instructional Leadership. *Phi Delta Kappan* 77, 380-81.

Joyce, B., and Showers, B. (1996). Staff Development as a Comprehensive Service Organization. *Journal of Staff Development* 17(1), 2-6.

Kleinsasser, A. (1999) Personal communication to J. Stepans.

Kleinsasser, A., and Miller, P. (1994) *An Evaluation of the TRIAD.* University of Wyoming Monograph. Laramie: University of Wyoming.

Leonard, W. (1997) *Methods of Effective Teaching and Course Management for University and College Science Teachers: How Do College Students Learns Science?* Dubuque, IA: Kendall/Hunt.

Loucks-Horsley, S., Harding, C. K., Arbuvkle, M. A., Murray, L. B., Dubea, C., and Williams, M. K. (1987) *Continuing to Learn: A Guidebook for Teacher Development.* Andover, MA/Oxford, OH: The Regional Laboratory for Educational Improvement of the Northeast and Islands/National Staff Development Council.

Melear, C. T., Hickok, L. G., Goodlaxson, J. D., and Warne, T. R. (1998) Responses of Preservice Secondary Science Teachers to Learning Science in an Apprenticeship: The Research Experience (RE). In *Translating and Using Research for Improving Teacher Education in Science and Mathematics,* edited by J. B. Robinson and R. E. Yager, SALISH II.

National Committee of Inquiry into Higher Education. (1997) *Higher Education in the Learning Society: Report of National Committee* (Dearing Report). Norwich: HMSO.

National Research Council (NRC). (1996) *National Science Education Standards.* Washington, DC: National Academy Press.

National Research Council (NRC). (1996) *From Analysis to Action: Undergraduate Education in Science, Mathematics, Engineering, and Technology.* Report. Washington, DC: National Academy Press.

National Science Foundation (NSF). (1999) National Science Foundation Chautauqua Short Courses. (*www.engrng.pitt.edu/~chautauq*).

Piaget, J. (1971) Problems of Equilibration. Piaget and Inhelder: On Equilibration. *Symposium of the Jean Piaget Society,* edited by C. F. Nodine, J. M. Gallagher, and R. D. Humphreys. Philadelphia: The Jean Piaget Society.

Rutherford, J., and Ahlgren, A. (1990). *Science for all Americans.* New York: Oxford University Press.

Saigo, B. W. (1999) A Study to Compare Traditional and Constructivism-Based Instruction of a High School Biology Unit on Biosystematics. Doctoral diss., University of Iowa.

Siebert, E. D., Caprio, M. W., and Lyda, C. M. (1997) *Methods of Effective Teaching and Course Management for University and College Science Teachers.* (Appendix B) New York: Kendall/Hunt.

Stepans, J. I., and Saigo, B. W. (1993) Barriers Which May Keep Teachers from Implementing What We Know about Identifying and Dealing with Students' Science and Mathematics Misconceptions. In

*Third International Seminar on Misconceptions and Educational Strategies in Science and Mathematics,* edited by J. Novak. Ithaca, NY: Cornell University.

Stepans, J. I., McClurg, P. A., and Beiswenger, R. E. (1995) A Teacher Education Program in Elementary Science That Connects Content, Methods, Practicum, and Student Teaching. *Journal of Science Teacher Education* 6(3), 158-63.

Stepans, J. I., Saigo, B. W. and Ebert, C. (1999) *Changing the Classroom from Within: Partnership, Collegiality, Constructivism.* 2nd ed. Montgomery, AL: Saiwood Publications.

Swanson, J. (1995) Systemic Reform in the Professionalism of Educators. *Phi Delta Kappan* 77(1), 36-39.

Sykes, G. (1996) Reform of and as Professional Development. *Phi Delta Kappan* 77(7), 465-67.

Wohlstetter, P. (1995) Getting School-Based Management Right: What Works and What Doesn't. *Phi Delta Kappan* 77(1), 22-26.

Wright, J. C., Millar, S. B., Kosciuk, S. A., Penberthy, D. L., Williams, P. H., and Wampbold, B. E. (1998) A Novel Strategy for Assessing the Effects of Curriculum Reform on Student Competence. *Journal of Chemistry Education* (Aug.).

Yager, R. E. (1993) The Iowa Chautauqua Program: A Model for In-Service Science Education. *Science Education International* 4(1), 26-27.

Yager, R. E., and Blunk, C. K. (1991) The Iowa Chautauqua Program. *Iowa Educational Leadership* 7(2), 65-69.

# Assessment Standards

Judith E. Heady, Brian P. Coppola, and Lynda C. Titterington

**A**ssessment is an integral link between effective learning and effective teaching. It is a measurement of student outcomes; however, it can be much more than an evaluation that leads to a grade in a course or on an assignment. In fact, when educators describe assessment they distinguish it from evaluation; as discussed in the *National Science Education Standards* (NRC 1996), evaluation is the end result of assessment. We are urged to assess our students' learning continuously and in many different ways, *then* to make judgments and provide a narrative or a grade to inform each student (that is, to evaluate them). The role of assessment is to show reliable evidence that learning relative to course goals has occurred. The *Standards* are congruent with this role because they emphasize the measurement of learning outcomes over mere course completion (Schneider and Shoenberg 1999; Cox 1995).

In addition to measuring outcomes, assessment can be used in at least two other ways (Brunkhorst 1996). First, assessment can be used to inform instruction. By studying the students in the class and collecting data that inform us about their progress, we can make a determination about the effectiveness of our teaching. Such data might lead to changes or might provide confirmation that our teaching approaches are effective. Second, a well-designed assessment can be used as a learning tool by students. Assessments that come early and often during the formative stages of learning provide students with feedback on the efficacy of their study habits and allow students to assess their own learning (Angelo and Cross 1993).

This chapter provides a consensus framework of criteria for the design of assessment tools and interpretation of resulting data. The *Standards* require that assessments have clearly stated purposes, a design that fits what needs to be measured, actions that are based on data, and procedures that are based on valid interpretation of data:

- Assessments must be consistent with the decisions they are designed to inform.
- Achievement and opportunity to learn science must be assessed.

● The technical quality of the data collected is well matched to the decisions and actions taken on the basis of their interpretation.

● Assessment practices must be fair.

● The inferences made from assessments about student achievement and opportunity to learn must be sound.

*Source:* National Research Council. (1996) *National Science Education Standards.* Washington, DC: National Academy Press, 78, 79, 83, 85, 86.

While this framework of criteria in the *Standards* was written for grades K–12, many individuals and groups in universities and colleges also have attempted to define assessment. For example, dialogues in the American Association for Higher Education's (AAHE) publications have resulted in the following working definition:

Assessment is an ongoing process aimed at understanding and improving student learning. It involves making our expectations explicit and public; setting appropriate criteria and high standards for learning quality; systematically gathering, analyzing, and interpreting evidence to determine how well performance matched those expectations and standards; and using the resulting information to document, explain, and improve performance. When it is embedded effectively within larger institutional systems, assessment can help us focus our collective attention, examine our assumptions, and create a shared academic culture dedicated to assuring and improving the quality of higher education. (Angelo 1995, 7)

This definition would support all levels of education with a minor change of wording in the last line, and it is inclusive for all disciplines including science.

In 1992, a collaborative effort of faculty from twelve colleges and universities resulted in a document entitled *Principles of Good Practice for Assessing Student Learning* (AAHE 1992). The nine principles of good practice for assessing student learning articulated in the document, which are consistent with the *National Science Education Standards*, reinforce and help to explain the five assessment standards of the *Standards* with respect to the postsecondary level. For example, the first principle of good practice centers on the educational mission of an institution with the emphasis on the improvement of what is valued most for student learning; another principle focuses on clearly stated goals. Yet another principle of practical importance in assessment speaks to "issues of use" so that the information gathered focuses on what is valued and whether it will actually provide for improvement. Importantly, the need for assessment is recognized to be part of a larger arena for improvement in teaching and learning and all decision-making.

The AAHE document recognizes that outcomes are important: They define what we should be measuring. However, because learning is a highly complex activity, it must be measured in a variety of ways. We must take great care in our zeal to measure learning that we do not adopt standardized tests that emphasize test-taking prowess or measure shallow, superficial knowledge (Sacks 1997). The forms of assessment should be appropriate for the practice of science and be varied enough to be able to measure a level of performance reliably. The AAHE document also notes that each step in student learning must be measured; assessment needs to be continuous in order to see progress. Moreover, students need to have the opportunity to demonstrate their skills and knowledge over time so that daily variations in student performance, which inflict bias into the data, are minimized.

Why do we need to think about the *Standards* for assessment at the postsecondary level? First, we should be striving for a seamless assessment system that continues to raise our expectations of learners and the expectations of learners themselves, from K to 16 and beyond. The public and its elected representatives are calling for more common measures of the effectiveness of higher education. Perhaps for too long the college instructor has been the examiner as well as the teacher without the presence of an agreed-upon set of national standards (Holyer 1998). The *Standards* afford scientists in higher education the opportunity to validate their expectations based on a professionally agreed-upon set of criteria. Second, essentially all of our students will be teachers in some way or another as adults—especially those who become parents. Therefore, we need to model sound educational practices, including ongoing assessment, in our classrooms. At the postsecondary level it may be helpful to talk with our students about assessment: what we are measuring, why we need to do that measuring, and what the acceptable standards are in the course. Third, an inference can be drawn from the scholarship of teaching and learning (Boyer 1990; Glassick, Huber, and Maeroff 1997). Shulman's leadership initiatives at the Carnegie Foundation for the Advancement of Teaching are devoted to improving how we think about teaching and to learning how to work with the emerging modes of documentation for evidence of teaching and learning (Anderson 1993; Shulman 1993). From reading the foundation's work, we conclude that assessment of student learning can provide the kind of information we currently lack about the actual effects of a contemporary education. For example, assessment can

- inspire people to act and plan more thoughtfully,
- inspire more public discourse,
- promote multidisciplinary conversations and projects,
- recognize the need for greater professional development in future faculty,
- provide feedback information for personal improvement by reflective practitioners, and
- provide feedback to students, curriculum designers, and decision makers.

Finally, one should not consider assessment without linking it to teaching and learning. For example, Teaching Standard C (as discussed in Chapter 1) addresses the necessity of classroom assessment for both students and the instructor. Equally important, assessment cannot be considered without linking it to the curriculum content (as will be discussed in Chapter 4). These three aspects of education—teaching (pedagogy), curriculum, and assessment—must be aligned to achieve effective learning experiences.

This chapter offers some interpretations of assessment criteria and several vignettes that might act as guides; we show that the design and use of effective assessments is integral to any course and curriculum. We expect that readers will modify these guides to suit their situations.

## Coordination of Assessment with Intended Purposes

Assessments must be consistent with the decisions they are designed to inform.

● Assessments are deliberately designed.

● Assessments have explicitly stated purposes.

● The relationship between the decisions and the data is clear.

● Assessment procedures are internally consistent.

*Source:* National Research Council. (1996) *National Science Education Standards.* Washington, DC: National Academy Press, 78.

Assessment Standard A cites the critical features of all valid assessments. The purposes of an assessment must be clearly articulated; moreover, the experimental design for classroom research must be validated to confirm that the data collected support subsequent actions. In an assessment, the kind of information one asks for, when it is asked for, and how it is asked for all govern the ways in which it can be used.

Typical purposes of course assessment by teachers at all levels include the following: to give insight into student understanding; to provide a measure of progress toward course or instructional goals; and to assign a grade. When an evaluation comes at the end of a course or assignment, it is *summative*; summative evaluation generally results in rankings of student performance (e.g., grades) and certifies a level of competence against some standard. When an evaluation is *formative*, it is assessment that helps faculty to make changes in their teaching, if necessary, and gives students continual feedback on their performance; practice quizzes and multiple drafts of papers are examples of formative assessment. Both summative evaluation and formative assessments are important in teaching. Ideally, instruction should be peppered with opportunities for both the students and the faculty to learn whether the understanding between them is progressing effectively.

To obtain valid data, the use of several types of assessment tools is recommended. Multiple perspectives not only satisfy multiple goals, but also deliver complementary information. Like travelers who locate themselves with radio beacons, faculty and students alike are said to triangulate their understanding by using multiple sources of information to reach their conclusions. Triangulation allows instructors to take a single-dimensional result—such as student performance on a multiple-choice examination—and couple it with student interviews, classroom observations, or a task requiring an explanation in order to understand if students have developed the skills indicated by their examination scores.

Every course should have a well-articulated set of desired student outcomes. Effective teaching facilitates student progress toward achieving those outcomes; course goals, then, are based on outcomes (course design) and teaching (delivery). Assessing these goals across multiple dimensions allows us to better understand teaching and learning; assessment data and any such trends should align with the course goals. For example, faculty and students might assume that higher examination scores are correlated with a deeper understanding of the subject and a corresponding ability to make meaning from new information. By specifically collecting different kinds of evidence (interviews, written and oral tasks, observations, background attributes), faculty can test their assumptions about the relations among instruction, examination performance, and student learning. Alignment, or congruence, between instructional goals and assessment results is important; when this information is *not* aligned (e.g., student performance is linked to prior knowledge and not to the course), then instructors need to rethink their course design and instruction.

The following six categories summarize the types of assessment tools that are in relatively common use in higher education. Note that the tools in each category are complementary, differing in their intent and in the fundamental information they provide. The first three categories constitute traditional classroom techniques used by the faculty member or instructor in a course. The second three categories are composed of tools that can also be used by science faculty; however, these methods are most often used in research projects on how learning occurs. The interpretation of data from those projects requires a good understanding of cognitive or education science.

## Category 1: Examinations, papers, reports, and projects

These assessments are strictly product-based. We can only infer something about the process by which these materials are constructed, because the artifacts that indicate a student's process are missing. Although we may be satisfied that a student's ability to replicate an adequate performance is acceptable, it is incorrect to conclude that the appearance of any product implies a unique or even correct path. What appears to be a correct or acceptable presentation is primarily an interpretation of the instructor.

## Category 2: Periodic sampling of intermediate materials

In courses where the development of student learning is monitored, "drafts" or "studio practice" are often used. Writers of all kinds use multiple drafts to help sort out intermediate ideas. Nearly all of the so-called active learning strategies (or "classroom research") are ways for instructors to get feedback on the intermediate learning of students (Angelo and Cross 1993). These assessments treat the classroom setting as a kind of extended conversation, in which the instructor is asking the questions "Do you understand what I am saying?" and "Can you tell me how you know?" and the students are replying through various means. The evaluation of these assessments includes path as well as product. Portfolios are another common way to present assessment in this category.

## Category 3: Peer-based editing

When we create opportunities for structured peer review and critique—and teach the criteria for assessment and evaluation along with the subject matter—both the student being assessed and the student assessor benefit. Students improve their ability to assess their own work by examining how their peers approach the same assignment or task. Assessments in this category actually represent teaching events, in which the assessor is called on to express understanding—that is, the learner becomes a teacher. In fact, people learn quite differently when they are explicitly aware that they are going to have to teach what they are learning. In this sense, peer assessments are a fundamental reversal in the student-teacher roles (Coleman 1997; Coleman, Brown, and Rivkin 1998).

## Category 4: Performance-based assessment

Under the supervision of a facilitator, monitor, or recorder, students participate in a task that evokes an aspect of learning from the course. For example, laboratory students might be faced with an unknown substance and be asked to determine its identity. In performance-based assessment, students might be required to think aloud about how they would learn what the substance is, so that the strategic process used to solve the problem can be evaluated. Many expert-novice research studies on learning involve performance-based assessment. One means involves the videotaped responses of a subject group of undergraduates; the responses are coded and analyzed to compare the problem-solving strategies of the students with those of faculty and graduate students.

## Category 5: Large-scale survey work

Instructors may give pre- and posttests in a course to determine what learning has occurred. In this case, differences observed between pre- and posttest designs have to be carefully controlled for preexisting variations such as prior experiences of the student being tested. When the multiple responses on the posttest show statistically significant improvement over the pretest responses, the learning activity is judged successful; however, the difference between pre- and posttest performance may be influenced by a number of external factors such as differences in testing conditions. In interpreting the results, the researcher should make sure that the observed changes are self-consistent within the theoretical framework and that they make sense within the context of the course or intervention in question. When responses are self-reported, the statistical analysis and its interpretation may be quite complicated.

## Category 6: Interviews, observations, and focus groups

Rich anthropological studies can reveal important information about student learning and faculty instruction. These are term-long and yearlong studies of change that rely on direct interaction with students and faculty by a well-trained third party, who observes the process of instruction and learning and interviews the participants. Not surprisingly, such intensive studies are rare. There are some small-scale efforts in this category of Small Group Instructional Diagnosis (SGID), which include peer review by classroom visitation and one-time, midterm focus groups conducted by an external evaluator (Redmond and Clark 1982; Bennett 1987; Coffman 1991).

The following two vignettes describe assessments that are used, respectively, to develop self-assessment skills of students and to measure whether computational methods are integrated into students' analytical skills.

### From the Field

**Structured Study Groups (SSGs): Using Peer-Facilitated Instruction to Develop Self-Assessment Skills**
Brian P. Coppola, Department of Chemistry
University of Michigan, Ann Arbor

Since 1994, a cohort of 120 first-year chemistry students has earned honors credit within the 1,200-student course of standard work and examinations instead of in a separate section of the course. In the Structured Study Group (SSG) program, these students participate in extra, weekly two-hour sessions that are shaped, metaphorically, along the lines of a "performance studio" in the arts. Assignments, in the form of common (not identical!) tasks, are subjected to peer presentation and peer critique facilitated by upper-level undergraduate leaders. Students are not simply directed to work in groups or provided with problem sets, although both activities can be productive and engaging. Instead, students in the structured study groups follow a detailed curriculum that helps them to develop the kind of skills that I and my colleagues in the chemistry department believe are attached to a deep mastery of the subject. The format of the SSG sessions also encourages the students to develop their general learning skills, especially self-assessment.

During each session, the meeting time is typically divided among a number of activities. Each participant brings a duplicate set of his or her written assignment from the previous week. These assignments generally involve the creation of examples within a given context. In the very first assignment, they pick a C10-C13 molecule from a chemistry journal and are directed to construct five rational examples of molecules with the same formula (after learning, in their session, how to decode line formulas, what journals are, where they are found, and what proper citation format looks like). They then propose rankings for their created molecules based on any three properties—for example, magnitude of dipole moment, boiling point, and solubility—and write out their rationales. We direct the students to provide a brief statement that puts the information in context, a copy of the journal pages from which the example is derived, and a properly formatted citation. At the beginning of the SSG session, the students submit one copy of their work to their leader and dis-

tribute the other copies to the group. One or two rounds of peer review follow. The peer reviewers do not correct the students' paper, but rather answer "yes" or "no" to a set of questions about the work:

- Does the molecule or reaction fit the prescribed criteria?
- Is the format and information appropriate to the level of the class?
- Is the citation formatted correctly?

The discussion that follows is unstructured and generally lasts longer than might be expected, given the yes/no answers. The students learn through review of others' assignments and transfer that understanding to a critical review of their own work. All students have a structured opportunity to make, recognize, and correct their errors before they get to an examination. After the reviewing is completed, the reviews and the unmarked papers are returned to the originator, and he or she has a chance to decide if any corrections are needed. This second set of assignments and the reviews are collected, and they form part of the basis for the leader's evaluation of the student's performance that day.

At the end of the class period, the next week's assignment is presented, along with any supporting discussion, examples, or software training (in ChemDraw, Chem3D, and molecular modeling packages such as CAChe (Computer-Aided Chemistry, software from Oxford Molecular) that is needed to do the assignment.

We use a scanner-computer-projection system in class so that a student's hand-drawn or on-disk answers can be used as the basis of a group discussion, if it is appropriate. Each term of the course in which structured study groups are used ends with a multi-session project. In the first-term course, the students prepare and refine questions about short research articles written by local or visiting faculty members. On the last week of classes, the entire cohort of study group students meets with the author, who then fields the questions asked by student representatives. To end the second semester course, the students spend the last three weeks creating, refining, and peer editing their own case studies in scientific and professional ethics.

The honors students are graded for their participation in the weekly groups within the context of the larger 1,200-student course. Every week during the term, the group leaders and a faculty member meet to discuss the upcoming and previous assignments, the grading criteria, and the classroom challenges faced by the leaders themselves. The leaders are then responsible for assigning each student a grade of U (unsatisfactory), S (satisfactory), or O (outstanding). In electing to participate in the honors groups, students agree to have their course grades based on a two-part scheme. First, the entire class of honors and nonhonors students have their grades determined as usual, based on four examinations. For an Honors student to maintain this grade with an "H" designation, he or she needs to have achieved an S average or greater from the group leader, with an O counterbalancing a U. A less than S average results in a proportional reduction of the student's grade, with an all U average reducing the student's course point total by 10 percent.

## From the Field

### The Counterintuitive Event: A Performance-Based Assessment

Brian P. Coppola, Department of Chemistry
University of Michigan, Ann Arbor

In our second-term honors chemistry program, all of the students learn the CAChe (Computer-Aided Chemistry) program and apply it to their laboratory problems. A question we wanted to answer was whether these students integrated computational methods into their toolbox of analytical skills. Would the selection and use of the CAChe software be judged spontaneously and appropriately in situations where analyses using pen-and-paper or physical molecular models might also be an option? To test this, we performed a study using an interview-based format. Three groups of subjects were devised: (1) a group of faculty and graduate student experts, (2) a group of students who learned and used the software, and (3) a comparable set of students who did not learn the software. Our subjects were presented with chemical information on which a prediction was solicited. The two student cohorts differed, to the best of our ability to determine, only in their participation in extra group work each week (see Structured Study Groups, SSGs, in the previous vignette). All of the students were a part of the same large lecture class for their formal course work. Because all of the students who learned CAChe were part of the SSGs, we also used this study as an opportunity to gather information about whether the problem-solving skills of the students were affected by participation in the group work.

Students in SSGs spend two hours per week with peer-facilitated work that requires extensive use of library and software resources. The actual group work emphasizes peer review and analysis, as well as the development of self-assessment skills. One of our expectations was that these students would begin to see that the "answers" they create are the *beginning* of meaningful discourse about chemistry, not the end.

In our study, we present our subjects with a two-page problem.

**First page**: Subjects are presented with a series of five trimethyl Group IV substituent groups [$(CH_3)_3X$-, $X=C$, Si, Ge, Sn, and Pub)] and asked to pre-

dict the order of relative energy difference between the two chair forms of the corresponding monosubstituted cyclohexane derivatives. The nature of the data is such that the most likely prediction ($Pb>Sn>Ge>Si>C$) will be the opposite of the experimental results ($C>Si>Ge>Sn>Pb$). In the presence of an interviewer, the responses of the subjects were tape recorded while they described their thought processes. Once a prediction was made and responses completed, the subjects were instructed to turn the page.

**Second Page**: After confronting the actual experimental results (presented as a graph on top of the second page), the subjects are instructed to reconcile the counterintuitive trend. Once that is completed, we ask our subjects what they would do to test their ideas. It is at this point that we want to discover whether or not computational chemistry has been added to the list (or "toolbox") of useful skills. Once the interviews are completed, the tapes are reviewed and coded.

The expert group demonstrated the following attributes:

- All of them began by restating the problem.
- All of them made a fairly early prediction after taking an inventory of the major factors related to the problem. This prediction was followed by a fairly extensive explanation.
- The thought process used by the experts was cyclical: examination of an alternative model, rejection

on the basis of a counter argument, and proposal of a new model.

- All of the experts relied on primary literature sources, the design of new experiments, and computational chemistry methods.

All of our student subjects were in the same large lecture class. The group of honors students within this course was divided, by self-selection, into those who elected to participate in SSGs and those who did not. Historically, the group of honors students scores 8 to 10 percent higher on their examinations than the nonhonors students. As a cohort, the examination performance of the non-SSG honors students was identical with their historical counterparts, equivalent to a random placement based on exam scores. The SSG students score an additional 6 to 8 percent higher than traditional honors students. We reasoned that weekly participation in correcting and revisiting one's own work would

translate to greater facility in reconciling a counterintuitive event.

The summary of how our two student cohorts fared against the attributes demonstrated by the expert group is shown in Table 3.1.

After confronting the actual results and reconciling the counterintuitive trend, we asked our subjects what they would do to test their ideas. All of the experts relied on primary literature sources, the design of new experiments, and computational chemistry methods. Our study indicated that our strategy for involving students with CAChe does indeed integrate the option of computational methods into the analysis of new chemical problems. Our findings, as we expected, extend beyond the assessment of the students' use of computational chemistry. In particular, we found that students who participate in their chemistry learning in a more structured discussion (SSG) format

**Table 3.1**

*Comparison of Approaches to New Chemical Information by Expert Group, Study Group Students, and Non–Study Group Students*

| Experts (Faculty and Graduate Students) | Study Group Students Who Used "Expert" Approach | Non-Study Group Students Who Used "Expert" Approach |
|---|---|---|
| Restated the problem | None | None |
| Made early predictions | All | None |
| Used cyclical analysis | Most | Few |
| Relied on primary literature, new experiments | All | None |
| Consulted textbook, TA, professor | Some | All |
| Used computational chemistry | Some | None |

more often developed an "on-their-feet" ability to think through the way they resolved a counterintuitive situation. The other group of students, for whom the only significant identifiable difference in experience was not having participated in SSGs, were inclined to try and resolve the problem by making a single unexamined proposition and then spend their time justifying their choice.

## Measuring Student Achievement and Opportunity to Learn

Achievement and opportunity to learn science must be assessed.

● Achievement data collected focus on the science content that is most important for students to learn.

● Opportunity-to-learn data collected focus on the most powerful indicators.

● Equal attention must be given to the assessment of opportunity to learn and to the assessment of student achievement.

*Source:* National Research Council. (1996) *National Science Education Standards.* Washington, DC: National Academy Press, 79.

Assessment Standard B speaks to assessment in science disciplines, and it is relevant for all educational levels. It directs instructors to focus on the science content that is most important. Assessments are one way of communicating to students what learning the teacher values in the course; thus the *Standards* require that we re-evaluate our assessments to be certain that they truly do "focus on the science content that is most important for students to learn." The specific content goals of the *Standards* (see Chapter 4) cover three major subject areas: physical science, life science, and Earth and space science.

In addition to knowledge of science facts, the Content Standards emphasize the importance of developing students' conceptual understanding. Thus, the Content Standards broaden the goals of assessment beyond a simple evaluation of content knowledge; assessments now must encompass "the ability to inquire,…the ability to reason scientifically, the ability to use science to make personal decisions and to take positions on societal issues, and the ability to communicate effectively about science" (NRC 1996, 79, 82). These statements call for a rigorous system of assessment that challenges students to move beyond simple recall and comprehension and into the realm of such higher-order thinking as application and synthesis. Students learn as they are tested (Elliot et al. 1996). If the test demands they remember facts, then they will remember as many items as they can. On the other hand, if the test requires understanding, creativity, and the ability to communicate science content, then students will also practice these skills.

To assess these important skills adequately, data on *multiple* aspects of student achievement need to be collected. Few paper-and-pencil tests assess the higher cognitive levels of thinking (Bloom et al. 1956); they usually fail to assess students' ability to transfer knowledge, to use knowledge as the basis for decision-making, or to communicate its significance to a peer. As K–16 science teaching methods expand to include more critical thinking and investigative techniques, assessment methods need to be adjusted to match these new teaching methods (Trowbridge and Bybee 1996). Some alternative assessments being used in university and college science courses include concept mapping; the use of journals, portfolios, and essays; and performance-based assessments that require students to complete extended projects, design experiments, or evaluate situations (Angelo and Cross 1993).

Assessment Standard B recognizes that assessment is closely related to instruction and the *opportunity to learn* as well as to student achievement. Opportunity to learn refers to the conditions that must exist for all students to learn science. These conditions include not only the materials and equipment available for student use but also, for example, the content and teaching knowledge of the instructor, the nature of classroom experiences as they relate to an understanding of how people learn, and the student-teacher contact time. Assessment of the extent to which these conditions exist should be ongoing and should serve to instigate change where appropriate. Students cannot be held accountable for achievement until professors provide adequate conditions for learning to take place.

In addition to measuring student achievement and opportunity to learn, the assessment *itself* can provide an opportunity for students to learn. Informal assessments in forms such as questioning, journals, and One-Minute papers (see pages 156–157) provide both the teacher and the student with feedback on how effectively the material is being taught and learned (Committee on Undergraduate Science Education 1997). Many of these alternative strategies can be incorporated easily into large lecture courses. For example, after discussing a particularly difficult topic, the instructor might pause and ask students to jot down the "take home message." Alternatively, a sample question can be presented to the class for students to discuss and answer. This system of assessment shows students that even failure can be an opportunity for learning—if students revisit the material, revise their answers, and learn from their mistakes.

The following vignettes describe the importance of formative assessment in learning and the measurement of student achievement meshed with the opportunity to learn.

# From the Field

## Using Journals to Assess Student Understanding of Anatomy and Physiology

Lynda C. Titterington, Biological and Physical Science Department
Columbus (Ohio) State Community College

I use student journals as a means of informal, formative assessment to help students in my community college anatomy and physiology classes. The journal serves a dual role in learning and assessment. Originally designed to help students learn and prepare for the departmental midterm examination, the study journal has evolved into a valuable tool that integrates learning and assessment. Not only do journals enable me to catch misconceptions and other problems before the examination, but they also provide students with a means of self-assessment. With the incorporation of an informal peer review process, students also have the opportunity to review material and process information more deeply (Woolfolk 1998).

I describe the study journal as a free-form, multipurpose record of students' work, which provides many varied opportunities for students to learn material. For example, the journal rewards students for their out-of-class study efforts, for reworking course material so that they can understand it, and for studying in small chunks to reduce last minute cramming. The journal also contains assignments designed to help students break down an overwhelming amount of material into more manageable bits, to relieve test anxiety by allowing students to practice for exams and to work at their own pace, and to encourage students to find opportunities to use course material in their daily lives. The theoretical basis for the journal as a learning and assessment tool is that it gives each student an opportunity to develop an individual learning style (Dunn, Beaudry, and Klavas 1989; Snow, Corno, and Jackson 1996) as well as to monitor self-understanding. The journal is particularly helpful for hands-on learners who learn by diagramming or sketching out processes for themselves. Students who try this strategy are often surprised at how efficient it can be.

Journals are also appropriate for adult learners who are already in the work force; these students are encouraged to write about their prior experiences or how they are already using anatomy and physiology in their work (Merriam and Caffarella 1991). It is a challenge, however, to convince adult students to try something different. Although integrating writing and science through journals is becoming common in elementary education (Reising 1997; Tierney 1996), few of my students have kept a journal outside of English class.

The study journal is worth fifty points (five points per week, and 10 percent of the total grade). The journals are graded by a simple rubric: three points for evidence of regular study and thought (three or more entries per week) and one point each for originality and organization. Instead of judging the content of each entry (Fulwiler 1987), the reader provides informative feedback such as comments and questions.

This system is designed to motivate students to study by rewarding them for effort (Clement 1987; Weiner 1994) and to give them a safe forum to experiment with new ideas and to learn from their mistakes.

The journals are similar to a portfolio in that I encourage the students to choose the types of entries they include. To get them started, the syllabus contains a list of ideas for the journals, such as Web sites to visit, problem sets from the textbook, and open-ended questions. Students are also invited to include a one-page summary of the lecture, to create their own diagrams, and to write their own review questions. Some of the students who are employed include relevant articles from trade journals and stories about how they are using anatomy and physiology in their jobs.

I use a peer assessment strategy for the journals, which has advantages for both the instructor and the students. As instructor, I am relieved of reviewing and commenting on each journal entry. For students, the peer assessment provides the opportunity to review course content as they review their classmates' work. Each journal contains a pair of assessment sheets

**Table 3.2**

*Assessment Sheet for Journal Evaluation*

| Week 1 | |
| --- | --- |
| *Content* | *Comments* |
| 3 = Three meaningful entries | |
| 2 = Two meaningful entries | I really liked |
| 1 = One entry or entries lack meaningful content | |
| *Style* | |
| 1 = Good reading!! | |
| 0 = Return to grammar school | |
| *Neatness and organization* | A suggestion for next week |
| 1 = No problems here | |
| 0 = Yikes! | |
| **1 Fascinating stuff!  Bonus point!!!* | |
| _____Total Points | Read by: _____ |

(Table 3.2); the assessment sheet includes both a rubric and space for commentary. The reader is required to make as many positive comments about the journal as possible and to include one suggestion for improvement. Every other week we have "journal day" in class and the students review two anonymous journals. I also look them over and provide a third score in cases where the two reviewers had vastly different marks.

The primary effect of the study journals has been to increase communication in the course. Since I have started using them, fewer students fail the examinations and those who do are more likely to approach me with requests for help. While only a few students have stated that they "enjoy" keeping a study journal, their comments indicate that they find the journals useful as a study guide. For example, many students insist on a journal day before an examination, "just to make sure I didn't miss anything." A typical student response to the journal is, "Overall it's a good idea. It can be a pain, but it really does help me keep up with the class."

## From the Field

### Teaching without Exams through the Use of Student-Generated Portfolios in an Undergraduate Environmental Geology Class

Kent S. Murray, Department of Natural Sciences
University of Michigan, Dearborn

Many of us at primarily undergraduate institutions have struggled with grading strategies in our courses. Early in my career I routinely used two or three hour-long exams and a final to assess student learning. In recent years however, science education debates have focused on the need for higher-order thinking, reasoning skills, and performance-based testing. Consequently, I have begun to rely more on collaborative and cooperative activities as tools for teaching geology (Macdonald and Bykerk-Kauffman 1995; Shea 1995). I use more group work and individual writing assignments (Wiswall and Srogi 1995), and I have incorporated the concept of mini-lectures given by students into several of my courses.

One method of assessment that I have found to be especially successful to accompany this approach to teaching, particularly in upper-division courses, is the use of student-generated portfolios (Slater and Astwood 1995; Astwood and Slater 1997). For the portfolio, students collect and present evidence that demonstrates that a particular subject has been thoroughly understood or "mastered." The portfolios represent a performance-based assessment such as that commonly used in the fine arts or architecture. Students must organize, synthesize, and clearly describe the information presented, where it was obtained, and how it is applicable to the subject at hand. In effect, the portfolio demonstrates that the student has significantly expanded his or her knowledge of the subject. Moreover, the portfolio must also include a paragraph of self-reflection that assures the reader that the student understands the relevance of the material and the processes of integration that have occurred during the learning process.

As discussed by Slater and Astwood (1995), a student-generated portfolio represents achievement relative to specific learning objectives that have been clearly stated at the beginning of the course. Because each portfolio is individualized, student assessment must be done by looking at the portfolio's contents relative to the course learning objectives. Each piece of evidence submitted in the portfolio is assigned a score of 0, 1, 2, or 3, based on the reader's judgment of the completeness of the evidence submitted and how representative the material is. A score of 0 or 1 suggests either that the evidence submitted was not appropriate for the particular subject or that no supporting discussion was included. A score of 2 may be awarded if the student has addressed the subject correctly—that is, presented acceptable evidence—but has discussed the subject only superficially and given little or no indication of having actually mastered the subject. Finally, a 3 is awarded when the student shows strong evidence of subject mastery by using facts to support opinions suffi-

ciently well to provide the reader with insight into the complexity of the subject. In addition, each piece of evidence in the portfolio that receives a 3 must include a "self-reflection" that clearly states why the evidence demonstrates mastery of the subject and why or how the mastery of the subject is important to the student.

During a typical academic semester of fifteen weeks, twelve different topics may be introduced and discussed in class along with specific case histories. These topics may range from geologic hazards such as earthquakes and landslides, to subjects that deal with the impact that people have on the environment, including soil and groundwater contamination. To demonstrate mastery of each of these subjects, students must investigate each of these topics. This can be accomplished by reading scientific journal, newspaper, and magazine articles; writing a research essay (minimum of three pages); conducting large-scale opinion surveys; investigating appropriate Web sites; taking field trips; or answering at least three end-of-the-chapter questions concerning that topic. To avoid submitting a portfolio that consists of all newspaper articles, each student must submit evidence from at least three journal articles and three newspaper or news magazines and must write at least three essays. The remaining three topics can come from whatever sources the student chooses. Whatever line of evidence is chosen, it is the students' responsibility to demonstrate clearly that they have mastered the objectives.

The portfolios are collected and graded three times during the semester. Typically the portfolios will represent 70 percent of the student's course grade, with grades assigned based on the rubric in Table 3.3. The remaining 30 percent will be derived from collaborative in-class exercises, short writing assignments, and mini-lectures.

Overall student response to a course without exams has been outstanding. Although there is still some anxiety associated with the portfolios, students realize that with careful design of their portfolios, they can learn and earn a grade they desire, which is not always the case with exams. One side benefit of the use of portfolios for the primary form of student assessment is that classroom discussions are generally livelier. Students are not concerned with taking copious notes and can spend more time involved in class discussions.

## Table 3.3

*Scoring Rubric for Student Portfolios*

| | |
|---|---|
| **A** | Strong evidence in at least 10 topics; adequate in other two |
| **B+** | Strong evidence in at least 10 topics; adequate in at least one other |
| **B** | Strong evidence in at least 9 topics; adequate in the others |
| **B-** | Strong evidence in at least 9 topics; adequate in at least one other |
| **C+** | Strong evidence in at least 8 topics; adequate in the others |
| **C** | Strong evidence in at least 8 topics; adequate in at least one other |
| **D+** | Adequate evidence in 11 topics |
| **E** | Adequate evidence in less than 9 topics |

Adapted from T.F. Slater and P.M. Astwood. 1995. Strategies for Using and Grading Undergraduate Student Portfolios in an Environmental Geology Couse. *Journal of Geoscience Education* 43 (3), 216-20.

# Matching Quality of Data to Consequences

The technical quality of the data collected is well matched to the decisions and actions taken on the basis of their interpretation.

● The feature that is claimed to be measured is actually measured.

● Assessment tasks are authentic.

● An individual student's performance is similar on two or more tasks that claim to measure the same aspect of student achievement.

● Students have adequate opportunity to demonstrate their achievements.

● Assessment tasks and methods of presenting them provide data that are sufficiently stable to lead to the same decisions if used at different times.

*Source:* National Research Council. (1996) *National Science Education Standards.* Washington, DC: National Academy Press, 83.

Assessment Standard C directs our careful attention to the match of data and consequences, along with using multiple assessments that are confirming and are authentic in relation to a discipline. Assessments need to allow all students over some length of time to demonstrate their understanding and ability to apply their knowledge.

This Standard duplicates some aspects of Assessment Standards A and B. Standard A also deals with purposes, collection of data, and decisions based on the data; Standard B also alludes to authentic assessment in that the focus must be on achievement data that measure what is most valued in the discipline. In all cases the use of multiple types of assessment is implied. In the discussion that follows, we do not reiterate these overlaps; instead, we look at the unique aspects of Standard C: the use of many assessments *over time* and the use of the data collected from the assessments.

To know what should be measured, we begin with clearly stated course goals. For example, the following goals might be part of a list for a course in which group research projects are assigned: to understand concepts related to a specific topic; to orally present relevant research for the group project in order to inform the other students; and to write a research paper that includes discussion of course concepts based on the group research project. Assessment Standard C says three important things about assessment of such a course:

**1** Sound inferences can be made only from data that is consistent over time. This requires that students be given opportunities to demonstrate progress toward the course goals in a variety of ways—for example, pen-and-paper tests, oral class presentations, writing assignments in the style of the discipline, and explanations of experimental procedures used to collect data for their projects. If a student consistently does well explaining the concepts in oral presentations and research writing but fails on pen-and-paper tests, the "student's performance is [not] similar on two or more tasks that claim to measure the same aspect" (NRC 1996, 83). Since the pen-and-paper test results are not well matched with the other assessments, it indicates that the data are not stable enough to evaluate student understanding fairly. It should also provide a caution flag for instructors who use *only* tests to evaluate their students' achievements. The positive outcome in this example is that two out of the three forms of assessment do show congruence. Clear and understandable rubrics for each assessment should provide students and instructors with reproducible data in order to obtain fair assessments of student efforts (Walvoord and Anderson 1998).

**2** Valid or sound assessment data are acquired over time. The data should generally show that the level of student achievement over time—in a course or as the student progresses through the program curriculum—is stable or improves with experience. The path of consistency or improvement ought to be evident over time if enough chances for demonstration are afforded. When only midterm and end-of-term evaluations are done, "bad days" cannot be corrected for. Assessing several written assignments that measure understanding of a specific concept from various perspectives should provide not only a consistent picture that confirms other types of assessment, but also a longitudinal view that strengthens the decision for the final evaluation or grade for the course.

**3** Data collected from classroom assessment include not only student grades and other achievements but also evaluation of the pedagogy and of the means of assessment. Inferences based on the data collected should be used to alter instructional strategies, if appropriate, and to prepare for the next offering of the course. One way to organize classroom assessment data is to keep a course journal or portfolio (Cerbin 1996). To decide if one's judgments about data collected from the course assessments are meaningful and consistent, the instructor can check validity and collect information from various sources that do confirm the conclusions.

The following two vignettes illustrate how assessment data collected during the formative stage of learning course material allow instructors to alter teaching techniques and time spent on task, as well as to document improvement in student understanding.

## From the Field

### Effective Use of Pretests and Posttests

Judith E. Heady, Department of Natural Sciences
University of Michigan, Dearborn

The classes I teach range from introductory biology to upper-level embryology and histology to a graduate course for teachers. In each class I administer a pretest and a posttest. The pretest tells me where my students are starting and often uncovers misconceptions that they bring to the course; the posttest provides a measure of their knowledge in that area of biology.

By examining the answers given on a pretest, I am able to fit the presentation to meet the needs of the students. As an example, when I see that most of the students understand what characteristics of an environment determine the sorts of organisms that are able to live there, I can spend more time on other topics or go into more depth in this area. If I discover a generally poor understanding of cell multiplication and basic genetics, I examine these topics with the students. Then I enlist the help of the experienced students to work with the inexperienced ones, and I can arrange to have some more difficult problems for those who are interested. The posttest tells me how effective the time spent in (and out) of class has been in increasing student knowledge. When the posttest reveals the extent to which students in my introductory class are able to explain or diagram why both parents must each have a recessive allele in order to have a recessive offspring, I know that the interactive classroom with its peer support and learning activities has been well worth the effort!

On the pretest in my graduate course for teachers, I usually ask for a concept map using a list of several terms about health, the environment, genes, bacteria, and population. Some students are very experienced in constructing concept maps, while others have little idea of what to do. Later in the course, after discussing most of the topics represented on the list, students practice doing a map alone or with others and then display them for the group. We follow this activity with a discussion about why the maps are helpful and how they might be used for assessment of understanding of a subject area. For the examination I generally provide a more comprehensive list of terms and ask for individual maps showing the relations among the

terms. In one offering of the course, all twenty-two students demonstrated increased learning by constructing excellent maps.

I teach my embryology class without a textbook, lectures, or the usual weekly laboratories; instead, the students engage in discussion and in term-long group research projects. Through the discussion of research articles, we place the concepts into the whole picture of animal development (Heady 1997). After a pretest during the first week, I review some basic cell biology before beginning the paper discussions. During the last week, students take a test that is very similar to the pretest. They generally show vast improvement in their understanding of embryology (Table 3.4); in fact, they use examples from their readings. While I used only three questions for the pre- and posttests in histology, I did see an improvement in understanding and explanation of the concepts.

All of my pretests and posttests have questions or problems that can be explained in more than one way. I ask about important concepts for which I want to see improved explanations and examples. I do not grade these tests, but I read them all and score the answers "correct," "part-correct," "incorrect," or "blank." Table 3.4 gives overall average percentages of correct answers from the classes.

The evidence indicates that effective learning and teaching have occurred. I know which of the concepts

## Table 3.4

*Comparisons between Pretests and Posttests in Four Courses (1997-98)*

| Course | Pretest* | Posttest* | Test Administration |
|---|---|---|---|
| Introductory Biology | 14 (2-32) | 69 (27-100) | Fall 97 (44**) 8 questions |
| Introductory Biology | 20 (0-62) | 87 (65-100) | Winter 98 (48**) 6 questions |
| Introductory Biology | 15 (3-32) | 69 (23-100) | Fall 98 (56**) 8 questions |
| Histology | 17 (5-33) | 77 (70-89) | Fall 96 (21**) 3 questions |
| Embryology | 27 (0-91) | 69 (32-95) | Winter 97 (19**) 12 questions |
| Biology for Teachers | 36 (0-95) | 87 (36-100) | Summer 97 (19**) 9 questions |

*Average percentage correct and range; does not count partially correct answers.
**Number of students taking pretests and posttests.

are more difficult to understand; I am also able to quantify the variation among students who start my classes, which helps me to use my class time more wisely. Because of the dramatic improvement in performance, students have responded favorably to the pre- and posttest exercise, and I have evidence that supports my feeling that my teaching has improved.

## From the Field

### Using Assessment in Curriculum Reform

Gail Schiffer, Ben Golden, and Gary Lewis
Department of Biological and Physical Sciences
Diane Willey, Department of Educational Research and Assessment
Kennesaw State University

Several years ago, we received a National Science Foundation Course and Curriculum grant (DUE # 9354758) to design and implement an interdisciplinary science sequence that uses student-centered learning techniques. Two primary goals of the sequence were to improve students' knowledge of science process skills and to improve their attitudes about science. One of our first tasks was finding appropriate assessment tools to measure progress toward both of these goals. After a thorough search of the literature, we selected the Student Attitude Inventory (SAI) developed by Richard Moore (Moore 1969; Moore and Sutman 1970) for assessing changes in science attitudes. This instrument measures acceptance of six beliefs central to scientific thinking.

Unfortunately, there were no instruments available for testing, at the college level, all of the process skills we wanted to assess. One of our criteria was that a test have two versions so that we could pre- and posttest all students. To answer this need, Willey and Schiffer wrote and validated two versions of the Science Process Skills Test (SCIPROS). In creating the SCIPROS Tests, we selected seven skills to assess: identifying hypotheses, identifying variables, measuring variables, identifying experimental design flaws, identifying and interpreting graphs, identifying assumptions, and identifying conclusions. We then wrote a series of everyday or general science scenarios and asked from one to four science skill questions about each. We wrote our questions so that they were free of specific science content, assuming only minimal science knowledge. For each skill tested, there were six questions that were evenly split between "life science" and "physical science." The questions were carefully arranged in the two versions so that similar questions (with respect to skill tested and kind of scenario) appeared in the same order. Scenarios were edited to avoid ethnic and gender bias. Validation results indicate that the two versions have sufficient reliability estimates to make them suitable for com-

paring curricula and programs.

Once we had our assessment instruments, we administered them to students taking the first science course in discipline-specific nonmajor sequences (biology and chemistry), in our interdisciplinary science sequence, and in the first introductory courses for majors (biology, chemistry, and physics). Students in all sequences were given a pretest, a test at the end of the first course, and a test at the end of the second course.

Examination of the attitude survey results indicated that, during the first course of the sequence, attitudes in the courses for majors and in the discipline-specific courses for nonmajors did not change, while those in the interdisciplinary course increased significantly. The data for the second course in the sequence are not as straightforward. During both years of the two-year study, attitude scores for students in the major courses remained unchanged while scores for discipline-specific nonmajor courses increased. For the interdisciplinary course, scores increased during the first year but not during the second year. The change in the pattern the second year is puzzling, but it may have been caused by a change to larger classes and the introduction of new teachers to the sequence.

The change in science process skills tells another story. While the SCIPROS scores of students in sequences for majors and discipline-specific sequences for nonmajors showed no real change over the sequence, those of students in the interdisciplinary sequence increased steadily and, by the end of the second course, were significantly greater than those of students in the science major sequences.

These results clearly indicated that the instruction in the integrated science courses was at least as effective in developing a more positive attitude as other nonmajor courses and that it was more effective at providing a greater understanding of science process skills than other introductory courses. These attitude and science process skill assessment data played important roles in curriculum reform at our university. Although many other factors also went into the decision, the university has established these interdisciplinary courses as the only nonmajor science sequence in our curriculum.

## Avoiding Bias

Assessment practices must be fair.

- Assessment tasks must be reviewed for the use of stereotypes, for assumptions that reflect the perspectives or experiences of a particular group, for language that might be offensive to a particular group, and for other features that might distract students from the intended task.

- Large-scale assessments must use statistical techniques to identify potential bias among subgroups.

- Assessment tasks must be appropriately modified to accommodate the needs of students with physical disabilities, learning disabilities, or limited English proficiency.

- Assessment tasks must be set in a variety of contexts, be engaging to students with different interests and experiences, and must not assume the perspective or experience of a particular gender, racial, or ethnic group.

*Source:* National Research Council. (1996) *National Science Education Standards.* Washington, DC: National Academy Press, 85.

The concept of fairness addressed in this standard extends to all disciplines and to all grade levels. Fairness aims to eliminate stereotypes and group-specific assumptions and to modify assessment tools for those who have any disabilities or limits, especially with language. Assessments should support the varieties of learning styles, experiences, and interests of the students so that group bias can be nearly eliminated. For this task, vigilance is needed.

The introduction of bias into an assessment can be subtle and unintentional. For example, there is a large body of research that supports the belief that even with current coeducation, girls and women do not have equal opportunity for a first-rate education (Belenky et al. 1986; Rosser 1990; AAUW 1992; Sadker et al. 1993/94;

Sadker and Sadker 1994; Rosser 1995; Diller et al. 1996). Among the documented practices that might affect the assessment of women in college-level science courses are the following: different expectations for women and men, devaluation of accomplishments of women, lack of supportive mentors, the hostile classroom, and even sexual harassment (Morgan 1996). If women taking a difficult math test, for example, especially women committed to the study of mathematics, feel a threat that they might confirm negative female math stereotypes, their performance might suffer, and their commitment to math might weaken (Steele 1997). Furthermore, if assessment is based on oral classroom participation and if women are silenced—for whatever the reason—the assessment is not a fair one (Sadker and Sadker 1994).

Remedies do exist, however. When spatial ability tests were changed from timed to untimed conditions, women students who had lower S.A.T. scores than the men students no longer scored significantly lower on these tests (Goldstein, Haldane, and Mitchell 1990). When highly motivated women math students were told either that there were or were not gender differences in the results of a challenging test, they performed worse or the same, respectively, as highly motivated men on the same exam.

Bias in assessment may also appear as a result of stereotyping racial groups. For example, African-Americans and other marginalized groups in our society have been excluded from opportunities because they have progressed through schooling being stereotyped as not doing as well as the majority (Jencks and Phillips 1998). Lowered expectations might be fulfilled and therefore close off opportunities for professional and graduate schools (Gose 1995). In fact, achievement differences measured by standardized tests for the most part show that gaps between whites and blacks increase as students move through K–12 (Mickelson and Smith 1991). Claude Steele (1997) has introduced a descriptive phrase, *stereotype threat*, to describe what happens even to the best and most committed students if they feel that the possible lack of ability to do well on a test might jeopardize their hard-won place in their intellectual world. He and his colleagues gave the same tests to the students who were told either that the tests measured ability or that they did not measure ability. In the first case white students outperformed black students and in the second case they had equal performances. The researchers were not describing test anxiety in these high-performing minority students, but were describing a situational frustration during a challenge that seems to trigger the fear of fulfilling societal expectations about them as part of some "group." Student performances on examinations such as the S.A.T. appear to be tied in part to some social influences such as gender and racial stereotypes.

Another example involved Asian-American women students who were reminded before taking a math test that they were "women" or were "Asian." Their scores were compared with those of other students who had not been told anything. In this case the first group had lower scores than the second group, indicating the fulfillment of negative female or positive Asian stereotypes (Shih, Pittinsky, and Ambady 1999). Perhaps some of the achievement differences among recognized gender and ethnic groups on the Medical National Board Part I Examination could be attributed to fear of fulfillment of stereotypical expectations among these highly committed students (Dawson et al. 1994).

When expectations are raised, when students are supported in their learning, and when measurements of learning are fair, the gap disappears (Rosenthal and Jacobson 1974; Rose 1995; Steele 1997; Basinger 1998). As an extension of this, public education at all levels must be supported equitably (Fruchter 1997; Kozol 1997).

Another group of students who have not been studied well are those with physical problems that might limit participation in science laboratories or field studies and hence affect fair assessment of their learning. Are we prepared to integrate these students into the total experiences of the course so that they are also prepared to be assessed fairly? Students who have difficulties reading and writing on examinations due particularly to language unfamiliarity might also need to have some compensation in their assessments. Sometimes the root of academic difficulty is not obvious. For example, a student left parts of problems unfinished on hourly examinations. Her difficulty was eventually attributed to a language problem and, given ample time on the final examination, she was able to demonstrate her true level of understanding.

One way to minimize bias is to engage the students themselves in developing assessments that will measure the instructor's stated outcomes; another way to minimize bias is to ensure that assessments are aligned with clearly stated course goals that students understand. Examples of ways to avoid bias in assessment are given in the following vignettes.

## From the Field

### Using Student Strengths to Develop Assessment Tools for Nonscience Majors

Suzanne Shaw Drummer, Department of Teaching and Learning
College of Education, Ohio State University

Sarah is the only science instructor at a small, four-year college primarily for art students. The students are not motivated to do well in science because they do not see its relevance to their lives. Moreover, many of the students do poorly in the class because the traditional teaching techniques used in science are not well matched to the learning styles of these predominately visual learners (Gardner 1991). Sarah wants to change her teaching style so that more of the students will have an opportunity to succeed at learning science. Because teaching and assessment are so integrally linked, she begins to examine her methods of assessing student learning.

The course that Sarah is teaching is an ecology course. She outlines the goals of her course, including the content and noncognitive concepts that she sees as desired course outcomes (Krathwohl, Bloom, and Masia 1964). One of her noncognitive goals is that students will become more aware of their own roles in affecting the quality of the environment. Sarah's challenge is to develop an assessment tool that will not only evaluate knowledge of course content but will also give her some insight into the learning occurring in the affective domain. Sarah also wants to make this assessment tool more suited to the learning styles of these primarily visual learners (Perrone 1994).

Sarah decides to have each student design a poster that either encourages or discourages some action affecting the environment. Each poster is to be accompanied by a one- or two-page position paper that explains the nature of the environmental problem, the student's personal attitudes about it, and a proposed solution. To make use of the students' expertise in art, Sarah uses a class period at the beginning of the semester to do brainstorming and small-group work to develop the assessment rubric for grading the project. In this way, all of the students have input into the grading standards and are aware of how they will be evaluated from the beginning of the project.

Similar poster assignments are required as part of each unit examination, with a topic being chosen from some area covered during that unit. The poster assignment counts for half of the grade for that unit; the other half comes from a more traditional type of assessment.

Students were enthusiastic when Sarah introduced the poster assignment. They worked effectively to develop an appropriate rubric for its scoring and included some artistic aspects that Sarah would not have thought of. The majority of the students performed well on these types of assignments throughout the semester. They were pleased to see an improvement in their grades and began to recognize that they could use their strength in art to improve their science understanding (and grades).

In addition, Sarah felt pleased with the result of her change in assessment. She began to spend more class time on discussions of environmental issues rather than just lecturing on the material (Shuell 1996). The students were more attentive and they participated in the discussions. Students learned to support their arguments in assignments, and the rubric allowed self-assessment of the finished product before it was turned in to the instructor.

In Sarah's situation the development of the assessment tool was relatively easy because her students were all art students with similar learning strengths. In a general science course, especially one for nonscience majors, there will be diverse students with many different learning strengths. In this case the instructor might consider an assignment "menu" from which students may choose the assignments that best fit their particular learning abilities (APA Board of Educational Affairs 1995). The lesson here is that science should be relevant and accessible to everyone. Research shows that more students learn more science when they see the relevance of science to their own lives and when their learning strengths are addressed (Woolfolk 1998).

## From the Field

### Literature-Based Examinations and Grading Them: Well Worth the Effort

Brian P. Coppola, Department of Chemistry
University of Michigan, Ann Arbor

Examinations, probably more than anything else, transmit our learning agenda to students. In our course, organic chemistry is structured so that state-of-the-art information from the primary literature used as case studies can be presented to novice students on examinations. This assures us that we are true to the facts of science and not simply inventing trivial derivatives of classroom examples. On the examinations, we include the citation along with some contextualizing statements. We are thus sending two messages to our students: (1) memorizing previous examples is not enough, and (2) understanding the subject matter of the introductory course allows the student to understand what chemists actually say about what they study. We reinforce the idea of multiple representations for the same phenomenon by asking students to provide words, pictures, graphs, and numerical versions of the same idea.

On nearly every exam, students suggest unanticipated but completely reasonable alternative solutions. These are important to note in class.

There are a few aspects of our grading practice that are worthy to note.

- **Make improvement count.** Because students develop their new skills at different rates, and because the course is truly cumulative, we have devised ways to make improvement count. One simple but effective technique is to increase the point value of exams throughout the term without increasing the length of the exam (three exams: 100, 120, and 140 points). It is worth more to do better later, so students receive the following messages: You do not have to be perfect at the outset, and practice has tangible value. It is likely that our students overestimate the modest mathematical value of this scheme. We give a final examination (240 points) in which we attempt to provide an even representation of the term. One way I gauge improvement is to compare these two measures of cumulative performance: the average of the three exams given during the term and the average on the final. Using a simple spreadsheet, I subtract the final exam average (as a percentage) from the average of points for the three exams (as a percentage). By ranking the class according to this difference, I can locate the subset of my class whom I call the horizontal performers: those whose difference falls between +8 and –8 percent (a range I have designated, through experience, as being within a reasonable variance for identifying a significant difference).

I first assign my grade breaks to the horizontal performers. I think that students improve as a result of skill building, so I look to the group above +8 percent as a separate group for grading. On the other hand, I think that students decline for enough different reasons that I cannot say for sure that it is due to loss of skills. Once I have set my grades for the horizontal group, I sort back in the decliners and award them the grade based solely on their scores, as determined by the grid of grades already established. However, I think that Student A, who accumulates 300 course points by scoring 20/100, 50/120, 50/140

(33 percent), and then 180/240 (75 percent), is fundamentally different from Student B, who scores 50/100, 60/120, 70/140 (50 percent) and then 120/240 (50 percent). Student B is a horizontal performer whose grade would correspond to whatever "50 percent" ended up being that term, while I want to recognize that Student A has improved. To assign a grade for these incliners, I use the final exam papers, the placement of their averages, and all due fairness to students with consistent performances. In this case, Student A would definitely receive a higher grade than Student B in my course, but not the grade earned by someone who scored at the 75th percent level consistently during the term.

- **Use an absolute scale.** Setting an absolute scale means more than saying that 90 to 100 percent is an A grade. Our system depends on the fact that we give common examinations and fundamentally agree on course standards. These standards were determined empirically. By the third year of Structure and Reactivity, we had enough experience to be able to set guidelines for performance based on correlating numerical values with the rich and informative student work presented in their papers. This system would not be easy with multiple-choice examinations. We have set our examination standard high, and we are comfortable with the idea that achievement above (or below) certain levels tells us about student performance (for more detail, see Coppola, Ege, and Lawton 1997).

● **Involve students in the process**. To grade examinations, we use graduate student instructors who are divided into groups. This technique has evolved over the years to a point where we can achieve a high level of consistency scoring over 1,000 exams with a staff of roughly forty graders in a four- or five-hour period. Since 1996, I have used a technique that attempts to demys-tify the grading process for my undergraduate students. During the grading session for the first examination, I look for two problems on which there are high variations in student responses. Before they are graded, I photocopy (or scan) the responses of four to six students. I then combine these into a handout with all identifiers to the originators removed. During class the next day, and prior to posting the exam key, I use the first twenty-five minutes to do an analysis of this handout. The total point values associated with the problems appear on the page. The students work in small groups to consider the answers to these problems and to create a fair grading scale, given the point values. This is, of course, exactly what I must have done prior to the grading session. After ten minutes, I call for the grading schemes and bring this discussion forward. Remarkably, the students will converge on the scheme that I created the previous evening within a point or two. This is an empowering experience. Even though the grading is already completed, my students get the sense that their thinking and my thinking converge. With the remaining time, I give the final grading scheme for these two problems and direct the groups to actually assign scores. Finally, I once again facilitate the feedback to the class by surveying the values assigned by members of the class to the work of their peers.

## Making Sound Inferences

The inferences made from assessments about student achievement and opportunity to learn must be sound.

● When making inferences from assessment data about student achievement and opportunity to learn science, explicit reference needs to be made to the assumptions on which the inferences are based.

*Source:* National Resource Council. (1996). *National Science Education Standards.* Washington, DC: National Academy Press, 86.

Assessment Standard E appears to state the obvious: (1) do not stretch the data to support unwarranted conclusions, and (2) clearly state any assumptions on which inferences are based. Too often data are used to support conclusions that are very loosely linked to data and without an understanding of conditions of the study. For example, factors that must be known when interpreting data include sample size, control or reference groups used, and external factors that have been controlled (and those which have not been controlled). Assessment Standard E may be regarded as a summary statement designed to keep everyone—teachers of science, administrators, and those who make policy decisions that affect science education—alert to the need for objective interpretation of assessment data. This requires looking at assumptions underlying the study and the data collected to make certain that the inferences on student achievement and opportunity to learn are valid.

At the college and university level, sound inference from assessment is feasible if assessments are carefully designed by faculty and related to the outcomes of the course and program. At Alverno College over the past twenty-five years, faculty have identified clear overall outcomes for college education. With a strong emphasis on collaborative work, it has been natural for faculty to discuss the meaning and interpretation of this standard in a collaborative way. The first vignette below illustrates how Assessment Standard E is reflected in the process used to design assessments in courses at Alverno. In the second vignette, the assumptions on which class assessment are based are included as a list of criteria, referred to as a heuristic.

## From the Field

### Assessment as Student Learning

Leona Truchan, Department of Biology
George Gurria and Lauralee Guilbault, Department of Chemistry
Alverno College

This essay addresses student learning and development through the design and implementation of assessment; our focus is assessment designed for assisting individual students. At Alverno College, an office of research and evaluation focuses on program evaluation. Results from assessments that faculty have developed have been used by this office to identify program and curricular effectiveness (a summary of the findings of the office of research and evaluation appears in *Assessing and Validating the Outcomes of College* [Mentkowski and Loacker 1985]). Our interpretation of student assessment is driven by a clearly defined framework, evolved by the faculty and informed by our best practice (Alverno College Faculty 1994). Designing an instrument based on these components is essential to make certain that assessment is truly a learning opportunity for the student.

Initially, we define the outcomes that the course instructor wants the student to demonstrate. These outcomes integrate student ability and course knowledge and are related to both the outcomes of the course and the program of study as a whole. For instance, each discipline has outcomes that ask students to use appropriate models to analyze and synthesize data. Specific criteria for each course relate to the stated outcomes. (For examples of outcomes and criteria for several courses, including an introductory science course and organic chemistry, botany, microbiology, and physics courses, see Truchan, Gurria, and Loacker 1997, 131–34. The major resource that explains Alverno College's assessment process is *Student Assessment-as-Learning at Alverno College* [Alverno College Faculty 1994].)

From these criteria, a specific "prompt" is created, ensuring that teaching, learning, and assessment are related. A "prompt" is a stimulus that elicits a student response or performance that embodies desired student abilities. For example, if the stimulus is the laboratory experiment performed in regard to renal function, the student might be given this situation: "As a senior at Alverno, you have been asked to give a speech on the scientific process of problem solving for later possible adaptation into a slide-tape presentation. You have chosen as a specific example the experiment in renal function done in the physiology lab. The presentation should be concise and interesting, of unspecified length. The audience is high school seniors considering college for a nursing degree."

Prompts, which are contextual and for a specific audience, bring students as close as possible to a personal and/or professional situation of learning within which they will be using the outcomes outside the classroom. The prompts and the ways they are carried out (modes) should be compatible with modes used in the discipline or subject and be congruent with best practice. For example, in the context of a simulated professional exhibit, the student prepares a short presentation, using media and demonstrations. He or she illustrates the theory of, practical uses for, and limitations of a new (to the students) technique or instrument.

The student should be able to demonstrate the outcomes in different ways, and the prompt should provide, for the student, a basis for self-assessment. After the prompt has been established, the criteria are made more specific so that they reflect not only the general aspects of the outcomes but also the more specific nature of the prompt and mode. The student demonstrates specific understanding of appropriate concepts. The criteria statements should be sufficiently clear so that they could be used by another assessor and understood by the student. Taken together, all of the criteria paint a picture of successful performance.

Criteria are used differently by students and by assessors and by students at various levels. Students should be able to take the set of criteria to make sound inferences and visualize what a successful performance is.

Based on this visualization, the student completes the assessment. The assessors use criteria as course standards, to which student performance is compared. And although beginning students tend to use criteria as directions for performance, more advanced students come to see criteria as a description of what a successful performance entails.

The ultimate goal of assessment is to teach the students how to identify what a successful performance is, and have them develop their own criteria to achieve this. The immediate goal of assessment is to infer student achievement.

## From the Field

### Assessment Criteria as a Heuristic for Developing Student Competency in Analysis and Evaluation of Published Papers in the Sciences

Robert A. Paoletti, Department of Biology
King's College

The volume of newly published research in the sciences, the rapidity with which such material continues to emerge, and the increasing availability and use of online sources of information make it imperative that college and university students learn to analyze and evaluate the literature.

To assess the developing competency of students in their analyses, I have developed a list of assessment criteria (Table 3.5) based on the scientific method paradigm. The criteria identify the important features of any scientific article and are derived from elements widely recognized as being essential for any valid scientific study (Gower 1997; Summers et al. 1998). In addition to their assessment func-

tion, the criteria provide guidelines by which students can anticipate my expectations.

The list of criteria provides students with a comprehensive and systematic approach that ensures identification and extraction of the important features of an article. It has been used successfully by students to evaluate published articles of varied levels of complexity in a variety of subdisciplines of biology and has helped to keep students mindful of critical components of research that may only be implied in print. The criteria provide another demonstration that assessment strategies can play a major role in the learning process.

## Table 3.5

*Criteria for Analyzing Published Research Articles*

*Competency/Outcome:* The student is able to use the scientific method paradigm as a heuristic for analyzing and evaluating the practice of science by others, as reflected in published papers, articles, or reports, according to the following criteria:

The Student:

1  Evaluates the relevance of an article or report to a specific need by perusing the abstract, introduction, and/or discussion sections.

2  Identifies and articulates facts and/or observations that require explanation.

3  Summarizes the products of the inductive reasoning process that were used to formulate the hypothesis that the paper addresses.

4  Determines and articulates the hypothesis being tested in the experiment/study.

5  Uses the deductive reasoning process to list predictions made from the hypothesis.

6  Identifies, describes, and evaluates the sampling techniques and experimental design used to test the hypothesis.

7  Describes and evaluates the experimental controls.

8  Identifies, describes, and evaluates specific techniques and instrumentation used in experimental tests of the hypothesis.

9  Summarizes the data/results of experimental tests of the hypothesis.

10  Identifies and evaluates a descriptive and/or inferential statistical treatment of the data to assess the certainty associated with the results.

11  Lists and evaluates interpretations made from the data/results.

12  Lists and evaluates the major conclusions of the experiment/study.

13  Determines and articulates support or refutation of the hypothesis by means of the null hypothesis.

14  Devises and describes further experiments that the data, results, interpretation, and conclusions may warrant to provide additional support for, or refutation of, the hypothesis.

15  Integrates the results, interpretations, and conclusions into an existing conceptual model or devises a conceptual model from the result.

16  Discusses any significant implications for or relationships to human concerns that the results and conclusions of the experiment/study indicate.

17  Collaborates with others to identify and apply available expertise, differing frames of reference, and different contexts to uncertainties in the experiment/study and reaffirms or modifies conclusions from the original analysis and evaluation.

18  Articulates the application of specific information contained in the article or report to his or her more comprehensive objectives.

# References

American Association for Higher Education (AAHE). (1992) *Principles of Good Practice for Assessing Student Learning.* Washington, DC: AAHE.

American Association of University Women (AAUW). (1992) *How Schools Shortchange Girls.* Washington, DC: The AAUW Education Foundation.

Alverno College Faculty. (1994) *Student Assessment-as-Learning at Alverno College.* Milwaukee, WI: Alverno College Institute.

Anderson, E. (Ed.) (1993) *Campus Use of the Teaching Portfolio: Twenty-Five Profiles.* Washington, DC: American Association of Higher Education.

Anderson, J. R. (1993) Problem Solving and Learning. *American Psychologist* 48, 35-44.

Angelo, T. A. (1995) Reassessing and Defining Assessment. *AAHE Bulletin* (Nov.), 7-9.

Angelo, T. A., and Cross, P. (1993) *Classroom Assessment Techniques: A Handbook for College Teachers* (2nd Ed.) San Francisco: Jossey-Bass.

APA Board of Educational Affairs. (1995) *Learner-Centered Psychological Principles: A Framework for School Redesign and Reform.* Washington, DC: American Psychological Association.

Astwood, P. M., and Slater, T. F. (1997) Effectiveness and Management of Portfolio Assessment in High Enrollment Courses. *Journal of Geoscience Education* 45 (3), 238-44.

Basinger, J. (1998) Ithaca Forges Ties with School in Harlem. *The Chronicle of Higher Education* July 3, A7.

Belenky, M. F., Clinchy, B. M., Goldberger, N. R., and Tarule, J. M. (1986) *Women's Ways of Knowing: The Development of Self, Voice, and Mind.* New York: Basic Books.

Bennett, W. E. 1987. Small Group Instructional Diagnosis: A Dialogic Approach to Instructional Improvement for Tenured Faculty. *Journal of Staff, Program, and Organization Development* 5 (3), 100-104

Bloom, B. S., Englehart, M. D., Frost, E. J., Hill, W. H. and Krathwohl, D. R. (1956) *A Taxonomy of Educational Objectives. Handbook 1: Cognitive Domain.* Vol. 1. New York: David McKay.

Boyer, E. L. (1990) *Scholarship Reconsidered: Priorities of the Professoriate.* Princeton, NJ: Carnegie Foundation for the Advancement of Teaching.

Brunkhorst, B. J. (1996) Assessing Student Learning in Undergraduate Geology Courses by Correlating Assessment with What We Want To Teach. *Journal of Geoscience Education* 44, 373-78.

Cerbin, W. (1996) Inventing a New Genre: The Course Portfolio at the University of Wisconsin-La Crosse. In *Making Teaching Community Property*, by Pat Hutchings, 52-56. Washington, DC: American Association for Higher Education.

Clement, S. L. (1987) Dual Marking System: Simple and Effective. *American Secondary Education* 8, 49-52.

Coffman, S. J. (1991) *College Teaching* 39, 80–82.

Coleman, E. B. (1997) Using Explanatory Knowledge During Collaborative Problem Solving in Science. *The Journal of the Learning Sciences* 6(4), 347-65.

Coleman, E. B., Brown, A. L., and Rivkin, I. D. (1998) The Effect of Instructional Explanations on Learning from Scientific Texts. *The Journal of the Learning Sciences* 7(3+4), 387-427.

Committee on Undergraduate Science Education. (1997) *Science Teaching Reconsidered: A Handbook.* Washington, DC: National Academy Press.

Coppola, B. P., Ege, S. N., and Lawton, R. G. (1997) The University of Michigan Undergraduate Chemistry Curriculum. 2. Instructional Strategies and Assessment. *Journal of Chemical Education* 74, 84-94.

Cox, J. W. (1995) Assessment from the Inside Out. *Issues & Inquiry in College Learning and Teaching* 17/18(2), 19-29.

Dawson, B., Iwamoto, C. K., Ross, L. P., Nungester, R. J., Swanson, D. B., and Volle, R. L. (1994) Performance on the National Board of Medical Examiners Part I Examination by Men and Women of Different Race and Ethnicity. *JAMA* 272(9), 674-679.

Diller, A., Houston, B., Morgan, K. P., and Ayim, M. (1996) *The Gender Question, Theory, Pedagogy, and Politics.* Boulder, CO: Westview Press.

Dunn, R., Beaudry, J. S., and Klavas, A. (1989) Survey of Research on Learning Styles. *Educational Leadership* 44(6), 55-63.

Elliot, S. N., Kratochwill, T. R., Littlefield, J. and Travers, J. F. (1996) *Educational Psychology: Effective Teaching, Effective Learning* (2nd Ed.) Dubuque, IA: Brown and Benchmark.

Fruchter, N. (1997) Saving Public Education 5. *The Nation,* Feb. 17, 21-23.

Fulwiler, T. (1987) Guidelines for Using Journals in School Settings. *The Journal Book.* Portsmouth, NH: Boynton/Cook.

Gardner, H. (1991) *The Unschooled Mind: How Children Think and How Schools Should Teach.* New York: Basic Books.

Glassick, C. E., Huber, M. T., and Maeroff, G. E. (1997) *Scholarship Assessed: Evaluation of the Professoriate.* San Francisco: Jossey-Bass.

Goldstein, D., Haldane, D., and Mitchell, C. (1990) Sex Differences in Visual-Spatial Ability: The Role of Performance Factors. *Memory & Cognition* 18(5), 546-50.

Gose, B. (1995) Test Scores and Stereotypes. *The Chronicle of Higher Education* Aug.18, A3.

Gower, B. (1997) *Scientific Method: A Historical and Philosophical Introduction.* New York: Routledge.

Heady, J. E. (1997) Testing Strategy. In *The Hidden Curriculum: Faculty-Made Tests in Science,* Vol. 2, edited by S. Tobias and J. Raphael, 32-33. New York: Plenum Press.

Holyer, R. (1998) The Road Not Taken. *Change* 30(5), 41-43.

Jencks, C., and Phillips, M. (1998) America's Next Achievement Test: Closing the Black-White Test Score Gap. *The American Prospect* 40(5), 44-53.

Kozol, J. (1997) Saving Public Education 1. *The Nation,* Feb 17, 16-17.

Krathwohl, D. R., Bloom, B S. and Masia, B. B. (1964) *Taxonomy of Educational Objectives. Handbook 2: Affective Domain.* New York: David McKay.

Macdonald, R. H., and Bykerk-Kauffman, A. (1995) Collaborative and Cooperative Activities as Tools for Teaching and Learning Geology. Report on NAGT/GSA Theme Session. *Journal of Geoscience Education* 43(4), 305.

Mentkowski, M., and Loacker, G. (1985) Assessing and Validating the Outcomes of College. In *Assessing Educational Outcomes, New Directions for Institutional Research,* edited by P. Ewell, 47-64. San Francisco: Jossey-Bass.

Merriam, S. B., and Caffarella, R. S. (1991) *Learning in Adulthood.* San Francisco: Jossey-Bass.

Mickelson, R. A., and Smith, S. S. (1991) Education and the Struggle against Race, Class, and Gender Inequality. In *Race, Class, and Gender, An Anthology*, edited by M. L. Andersen and P. H. Collins. Belmont, CA: Wadsworth.

Moore, R. W. (1969) The Development, Field Test, and Validation of the Scientific Attitude Inventory. Doctoral diss., Temple University.

Moore, R. W., and Sutman, F. X. (1970) The Development, Field Test, and Validation of an Inventory of Scientific Attitudes. *Journal of Research in Science Teaching* 7, 85-94.

Morgan, K. P. (1996) Describing the Emperor's New Clothes: Three Myths of Education (In-)Equity. In *The Gender Question in Education, Theory, Pedagogy, and Politics,* edited by A. Diller, B. Houston, K. P. Morgan, and M. Ayim. Boulder, CO: Westview Press.

National Research Council (NRC). (1996) *National Science Education Standards.* Washington, DC: National Academy Press.

Perrone, V. (1994) How to Engage Students in Learning. *Educational Leadership* 51(5), 11-13.

Redmond, M. V., and Clark, D. J. (1982) Small Group Instructional Diagnosis: A Practical Approach to Improving Teaching. *AAHE Bulletin* (Feb.), 310.

Reising, B. (1997) The Formative Assessment of Writing. *The Clearing House*, 71-72.

Rose, M. (1995) *Possible Lives.* Boston: Houghton Mifflin.

Rosenthal, R., and Jacobson, L. (1974) *Pygmalion in the Classroom: Teacher Expectations and Pupil's Intellectual Development.* New York: Holt, Rinehart and Winston.

Rosser, S. V. (1990) *Female-Friendly Science.* New York: Pergamon Press.

Rosser, S. V. (1995) *Teaching the Majority.* New York: Teachers College Press.

Sacks, P. (1997) Standardized Testing: Meritocracy's Crooked Yardstick. *Change* (M/A), 25-31.

Sadker, D., and Sadker, M. (1994) *Failing at Fairness.* New York: Charles Scribner's Sons.

Sadker, M., Sadker, D., Fox, L., and Salata, M. (1993/94) Gender Equity in the Classroom: The Unfinished Agenda, *The College Board Review* #170.

Schneider, G., and Shoenberg, R. (1999) Habits Hard to Break: How Persistent Features of Campus Life Frustrate Curricular Reform. *Change* (M/A), 30-35.

Seymour, E., and Hewitt, N. M. (1997) *Talking About Leaving: Why Undergraduates Leave the Sciences.* Boulder, CO: Westview Press.

Shea, J. H. (1995) Problems with Collaborative Learning. *Journal of Geoscience Education* 43(4), 306-308.

Shih, M., Pittinsky, T. L., and Ambady, N. (1999) Stereotype Susceptibility: Identity Salience and Shifts in Quantitative Performance. *Psychological Science* 10(1), 80-83.

Shuell, T. (1996) Teaching and Learning in a Classroom Context. *Handbook of Educational Psychology*, edited by D. C. Berliner and R. Calfee, 726-64. New York: Macmillan.

Shulman, L. S. 1993. Teaching as Community Property. *Change* 25(6), 6-7.

Slater, T. F., and Astwood, P. M. (1995) Strategies for Using and Grading Undergraduate Student Portfolios in an Environmental Geology Course. *Journal of Geoscience Education* 43(3), 216-20.

Snow, R. E., Corno, L., and Jackson, D. (1996) Individual Differences in Affective and Cognitive Function. In *Handbook of Educational Psychology,* edited by D. C. Berliner and R. Calfee, 243-310. New York: Macmillan.

Steele, C. M. (1997) A Threat in the Air: How Stereotypes Shape Intellectual Identity and Performance. *The American Psychologist* 52(6), 613-29.

Summers, R. L., Woodward, L. H., Sanders, D. Y. and Galli, R. L. (1998) Research Curriculum for Residents Based on the Structure of the Scientific Method. *Medical Teacher* 20(1), 35-37.

Tierney, B. (1996) *Write to Learn Science*. Arlington, VA: National Science Teachers Association.

Trowbridge, L. W., and Bybee, R. W. (1996) *Teaching Secondary School Science: Strategies for Developing Scientific Literacy*. (6th Ed.) Englewood Cliffs, NJ: Prentice-Hall.

Truchan, L. C., Gurria, G., and Loacker, G. (1997) Assessment: Connecting Teaching and Learning. In *Methods of Effective Teaching and Course Management for University and College Science Teachers*, edited by E. Siebert, M. Caprio, and C. Lyda, 123-34. Dubuque, IA: Kendall/Hunt.

Walvoord, B. E. F., and Anderson, V. J. (1998) *Effective Grading: A Tool for Learning and Assessment*. San Francisco: Jossey-Bass.

Weiner, B. (1994) Ability versus Effort Revisited; The Moral Determinants of Achievement Evaluation and Achievement as a Moral System. *Educational Psychologist* 29, 163-72.

Wiswall, C. G., and Srogi, L. (1995) Using Writing in Small Groups to Enhance Learning. *Journal of Geoscience Education* 43(4), 334-40.

Woolfolk, A. E. (1998) *Educational Psychology*. (7th Ed.) Needham Heights, MA: Allyn and Bacon.

# Content Standards

| Eleanor D. Siebert |

The essential context of all teaching is found in the content and skills to be taught. In the *National Science Education Standards* (NRC 1996), content is broadened beyond concepts and facts; content is meant to include inquiry skills involving the relationship of technology and science, the relevance of science in personal lives and decision-making, and an understanding of science as a human endeavor. The Content Standards begin by setting out unifying concepts and processes that give a framework for learning science content in grades K–12 and beyond.

As a result of activities in grades K–12, all students should develop understanding and abilities aligned with the following concepts and processes:
- Systems, order, and organization
- Evidence, models, and explanation
- Constancy, change, and measurement
- Evolution and equilibrium
- Form and function

*Source:* National Research Council. (1996) *National Science Education Standards.* Washington, DC: National Academy Press, 115.

These unifying concepts and processes provide for a coordination of science learning among grade levels and science disciplines that reinforces prior learning and permits the introduction of concepts of increasing complexity at higher grade levels. Additional aspects of the core material are articulated in the seven Content Standards:

Standard A: Science as Inquiry

Standard B: Physical Science

Standard C: Life Science

Standard D: Earth and Space Science

Standard E: Science and Technology

Standard F: Science in Personal and Social Perspectives

Standard G: History and Nature of Science

*Source:* National Research Council. (1996) *National Science Education Standards.* Washington, DC: National Academy Press, 105-108.

Science as inquiry is prioritized as Standard A, while Content Standards B, C, and D involve concepts associated with more or less traditional divisions among science disciplines. The content described in Standards B, C, and D lists relatively few concepts; however, a deeper presentation of the material and a deeper understanding by students is required. Further, the concepts are not identified in terms of theories known so well by name to scientists, but rather in terms of the ideas to be understood. For example, Standard B (Grades 9–12) does not ask that students know the atomic theory; rather, it states that students should "develop an understanding of the structure of atoms."

Programs at most colleges and universities include courses offered in the rather traditional content areas of the physical, life, and Earth and space sciences that will include coverage of the concepts indicated in the *Standards.* While the overall specific course content may vary, depending on the needs of the students' major program, Content Standards B, C, and D articulate those concepts that cross the sciences and are considered by scientists to be fundamental to understanding our natural world. In all K–12 science courses, the *Standards* require that science be taught as science is practiced—which begins by asking questions—and with attention to unifying themes.

You will find that the Content Standards of the *National Science Education Standards* are written to include *all* students. In past years, a large portion of elementary and secondary students has not had adequate access to science learning. In the lower grades, too often science content is not well integrated into the curriculum; and in high school science courses, students of color and those who are less affluent than many of their peers are not well represented. In general, we can say that scientific literacy for our citizenry has not been a goal of science programs or of our educational system.

Another point about the Content Standards is that they do not focus solely on students' achieving factual knowledge, but on their *understanding* of that knowledge and the *ability* of students to use it in their lives. This need for "understanding" and "ability" has called for a change in our ways of assessing whether students have reached the standards (see Chapter 3).

# Science as Inquiry

Science as inquiry reinforces the notion that science should be learned in the way that science is done. This standard extends across grade levels, K–12, because inquiry is at the center of all that we know; scientists will claim that the standard of inquiry extends through formal education during college and university years and beyond. University faculty should have a special understanding of this standard, because it emphasizes that science generally should be taught as it is practiced. The challenge is in implementing this standard in our college classrooms—especially those with large enrollments—in a way that enhances learning. It is a challenge that cannot be ignored, however, because it is critical to changing the way that science is taught at the university level. The key component to defining science involves asking questions and constructing knowledge based on observations. Moreover, any knowledge is tested by making predictions that are then compared to observations for verification.

As a result of activities in grades K–12, all students should develop

● abilities necessary to do scientific inquiry

● understanding about scientific inquiry

*Source:* National Research Council. (1996) *National Science Education Standards.* Washington, DC: National Academy Press, 121, 143, 173.

Abilities necessary to do scientific inquiry include the ability—or willingness—to formulate a question and the ability to make precise, unbiased observations. In addition, scientists require a knowledge base to interpret the data and, often, a measure of creativity in understanding how observations relate to the question being asked. If these are the abilities that all students should develop, then it is imperative that they be taught by inquiry (see Chapter 1) and assessed in a way that requires they demonstrate these skills (see Chapter 3).

By understanding scientific inquiry, students come to understand the nature of science. For example, students begin to understand that science is fraught with uncertainty—minimized by careful investigative work—and that scientific knowledge is based on observations that are subject to revision through rigorous and continued testing.

The implication of Content Standard A at the postsecondary level is that inquiry should play a major role in college and university science courses. This poses some challenges in large lecture courses (Lumsden 1997) but emphasizes the importance of laboratory or field experiences. A major role of the laboratory or practical experience is to foster inquiry; thus, work should move away from tightly directed experiments meant to confirm some obvious principle and toward a more research-oriented experience. Laboratories that (1) are open-ended, (2) include collaboration among students, and (3) extend over a reasonable period of time are more closely aligned with research projects carried out by scientists. The following vignette illustrates how research techniques, typical of an inquiry approach to learning, can be embedded in a laboratory course that includes a variety of experiments.

## From the Field

### A Research Approach to the General Chemistry Laboratory
Eleanor D. Siebert, Department of Physical Sciences and Mathematics
Mount St. Mary's College

For over ten years, Mount St. Mary's College has offered a course for honors students taking the second semester of General Chemistry. Our goal for the course is to provide students with an understanding of the world in which we live so that they might come to want to know more. This means helping students to look carefully at the world, to ask questions, and to relate information to the fundamental concepts that unite the sciences—essential habits of scientists who are engaged in research. Thus, research techniques need to be effectively integrated into the course.

The course that has evolved usually has an enrollment of twelve to eighteen students. While the course is not restricted to honors students at the college, students enrolled in the course have demonstrated an above-average facility with chemistry and, more importantly, have expressed an interest in laboratory studies. About four topics are selected for study each term, and each assignment is viewed as a small guided-research project. Although the topics selected for study

vary from year to year, the initial and final activities have become traditional in the course.

Students begin by developing a workshop that is a part of a citywide "Expanding Your Horizons" conference for middle school girls. Many workshop participants are from the inner city and from an ethnic or racial group currently underrepresented in the sciences—closely mirroring the Mount St. Mary's College students themselves. Following a general theme, the honors students organize and develop an interactive activity. In this way, the chemistry students learn about a subject, and their creativity is sparked as they work out a way to present it to others. Recent workshop topics have included chemistry "wizardry," household chemistry, chemistry in nature, the chemistry of color, and a workshop on polymer chemistry in which the students set up an exploration of polymers by simulating a visit to a shopping mall ("Polyrama Mall"). The workshop activity promotes enthusiasm among the college students and provides a course component of civic responsibility, a value that the college fosters.

Three to four weeks before the end of the term, students develop and carry out an individually formulated research question. As a means to focus student efforts, I suggest a general topic for study; topics over the past years have included plant pigments, natural dyes, atomic absorption spectroscopy, and environmental concerns. Students begin with library reading and reflection to formulate a question. After conferring with me, they design and carry out an experiment to answer their question(s); they are encouraged to use any instructional instrumentation in our laboratories that may provide relevant information. A formal report of the individual research project is required; some of the reports have been of such high quality that they have been published in the Mount St. Mary's College student research journal.

What makes the honors course unique is not just the inquiry- or research-oriented approach to problem solving, but the group dynamics that make problem solving most effective.

To have enough time to carry out the smaller research studies that compose the rest of the course, as well as to learn the importance of teamwork in science, the class works as a group. Together they pose questions and then divide each question into smaller ones, with students investigating different parts of the picture. The class data are pooled for analysis so that each student is able to see her data as a part of the whole. Communication skills are stressed because students give oral presentations of data and discuss results; formal writing is developed throughout the term.

I believe that engaging in research or inquiry approaches at the undergraduate introductory level of science courses provides valuable learning for students. For example, research is an approach to problem solving that requires a special and almost contradictory combination of qualities: imagination, yet caution; receptivity to new ideas, yet skepticism; thoroughness, yet an ability to conclude; and so on. A researcher is a student for life; what better lesson to take from the general chemistry laboratory experience?

# Subject Matter Content

The standards for physical science (Standard B), life science (Standard C), and Earth and space science (Standard D) are specific to grade levels K–4, 5–8, and 9–12. The tables below show how the concepts at each grade level complement and reinforce those introduced at prior grade levels. The *Standards* do not prescribe a curriculum; instead they describe general concepts that may be incorporated into local curricula. At the college and university level, it is useful to understand what our students have been taught so that we may continue to build on and enhance their understandings.

**Table 4.1**

*Content Standard B: Physical Science—Understandings That Students Should Develop in K–12 Activities*

| Grade Level | | |
| --- | --- | --- |
| K–4 | 5–8 | 9–12 |
| • Properties of objects and materials | • Properties and changes of properties in matter | • Structure of atoms |
| | | • Structure and properties of matter |
| | | • Chemical reactions |
| • Position and motion of objects | • Motions and forces | • Motions and forces |
| • Light, heat, electricity, and magnetism | • Transfer of energy and increase in disorder | • Conservation of energy |
| | | • Interactions of energy and matter |

*Source:* National Research Council. (1996) *National Science Education Standards.* Washington, DC: National Academy Press.

**Table 4.2**

*Content Standard C: Life Science—Understandings That Students Should Develop in K–12 Activities*

| Grade Level | | |
| --- | --- | --- |
| K–4 | 5–8 | 9–12 |
| • The characteristics of organisms | • Structure and function in living systems | • The cell |
| • Life cycles of organisms | • Reproduction and heredity | • Molecular basis of heredity |
| | • Regulation and behavior | • Biological evolution |
| • Organisms and environments | • Populations and ecosystems | • Interdependence of organisms |
| | • Diversity and adaptations of organisms | • Matter, energy, and organization in living systems |
| | | • Behavior of organisms |

*Source:* National Research Council. (1996) *National Science Education Standards.* Washington, DC: National Academy Press.

## Table 4.3

*Content Standard D: Earth and Space Science—Understandings That Students Should Develop in K–12 Activities*

### Grade Level

| K-4 | 5-8 | 9-12 |
|---|---|---|
| • Properties of Earth materials | • Structure of the Earth system | • Energy in the Earth system |
| | | • Geochemical cycles |
| • Objects in the sky | • Earth's history | • Origin and evolution of the Earth system |
| • Changes in Earth and sky | • Earth in the solar system | • Origin and evolution of the universe |

*Source:* National Research Council. (1996) *National Science Education Standards.* Washington, DC: National Academy Press.

Reading the content development outlined in Standards B, C, and D above, we realize again that students bring prior knowledge and understanding to our college and university science courses. They then build new understandings on those existing knowledge bases. This theory of education, *constructivism*, which was discussed in the Introduction, is the theory on which the *Standards* are based. It is important for instructors to understand the "baggage" that students bring to courses and to present material in a way that allows each student to complement, correct, and build on their individual knowledge bases. The following vignette illustrates how we might rethink our courses for nonscience majors in order to allow our students to build solid understandings of fundamental science content, to see connections across the disciplines, and to become scientifically literate.

## From the Field

### Rethinking the Content in Nonmajors' Science Courses
Mario W. Caprio, Department of Biology
Volunteer State Community College

In scientific literacy courses (typically those courses earmarked in college catalogs as being for nonscience majors), one rarely finds any content on the cutting edge of the discipline. All we ever expect students to learn about science are a few principles and concepts that have long since moved from the domain of pure science to the realm of what we call the *body of cultural knowledge*.

When students read *On the Origin of Species*, for example, they are learning very basic biology through a piece of nineteenth-century literature that could arguably have a place in the English curriculum; and although Mendelian genetics is fine biology, it is biology that is more than a century old. This is the kind of science that educated nonscientists know as much for its historical significance as for its scientific value. Science is something that we humans do, and we ultimately integrate the information we derive from our scientific endeavors with the whole of our knowledge to help us make sense of our complex world. Darwin, Shakespeare, Mendel, Melville, Newton, Aristotle, Einstein, Michelangelo—they and so many others are fundamental to our cultural heritage. When we emphasize to our students the *scientific* nature of their work, we suggest it has some specialized value that may have little relevance to the nonmajor's world, and we risk frightening the science phobic in the audience.

Am I suggesting that the work of Darwin, Mendel, Newton, and other scientists has somehow moved across campus and into the humanities division? Should we change what we call the subject matter in an introductory science course to give it a name other than *science*? Of course not. What I do think we need to do, however, is to be intellectually honest with our students

and ourselves: What we are exposing students to in our scientific literacy courses is that that part of the cultural fabric that originally came from scientific pursuits is now integral to the worldview of scientists and nonscientists alike. We need to recognize, and our students need to understand, that most of the topics we teach to nonmajors represent what is actually the scientific literacy component of a broader and more general *cultural literacy*.

Program Standard B (discussed in Chapter 5) says that science programs "should be…interesting and relevant to students' lives…and be connected with other school subjects." When we begin to think of the science topics in Content Standards B, C, and D as constituting more than just *scientific* literacy, when we can appreciate science as an integral part of the larger cultural framework of our lives, we begin to see the far-reaching implications of this Standard.

In his book *Science and Human Values*, Bronowski (1990) talks about a person needing both a scientific and humanistic perspective of the world in order to have an accurate sense of reality. Bronowski's book powerfully and beautifully supports the liberal arts tradition. Educated people know the value of this point of view, and yet educators routinely compartmentalize the disciplines, assuming that the synthesis of a more holistic view will somehow magically happen in our students sometime before graduation. Even in our nonmajors' courses, we usually cling to our specializations, rarely integrating the scientific and the humanistic, rarely risking the construction of knowledge in ourselves in the presence of our students.

A broad interpretation of the *National Science Education Standards* is that they allow us to define the "science program" for nonmajors (and maybe for science majors, too) in larger, cultural terms. Depending on their academic majors, most nonscience majors take one, two, or occasionally three natural science courses. With so little science in their academic lives, it is easy for these students to endure the hardship, the way an agnostic might endure a compulsory church service, and remain unchanged by the experience. Only if we can integrate their science experience with the mainstream of our students' lives and give it meaning and purpose will there be any possibility that they will retain and enlarge the lessons they repeat so well on our exams.

There are several ways of making that connection. Guest lecturers are one approach—for example, an attorney talking about DNA fingerprinting, a historian on the impacts of infectious diseases, or a mathematician on music theory. Science does not exist in a vacuum even though some science courses do.

We can also ask students to select a topic for a paper or classroom presentation that illustrates the relationship between a special interest they have and the science course they are taking. In my classes, there was the social worker whose report on Alzeimer's disease included videotaped interviews of clients in three different stages of the disease (yes, informed consent issues were properly addressed); a Native American who needed to learn more about glaucoma so she could better help with an education program on the reservation; and the business/journalism major who was interested in learning more about how science is funded.

I have also used a consciousness-raising exercise in which I give each collaborative group a newspaper from which they select articles that are not about science or medicine *per se* but that have a connection (however circuitous) to, say, biology, chemistry, or physics. With just a little time and not too much coaching from me, students start to make the connections: A story about new housing starts sparked a discussion of habitat destruction; an article reporting approval of a maintenance dredging project raised concerns about the food chain; and a local newspaper's ad for a sale on canned tuna triggered a lively argument that suggested that these students had read (though they hadn't!) Hardin's (1968) "Tragedy of the Commons," the classic article about how parts of the planet are not owned by anyone but are used by everyone, with adverse consequences (e.g., overfishing, whale harvesting, pollution). It soon became difficult to find an article that was *not* related to science!

# Science and Technology

Content Standard E recognizes the increasing role of technology in our lives—a role so important and pervasive that the distinction between science and technology often becomes blurred. Technology is enabled by science; however, technology refers to the application of science for individual and societal uses that are often unimagined by scientists. This Standard says that students should understand the relation between science and technology, a standard that is emphasized in almost the same form throughout K–12 grade levels, as the following table shows.

## TABLE

Content Standard E: Science and Technology

| What Students Should Develop | Grade Level | | |
|---|---|---|---|
| | K–4 | 5–8 | 9–12 |
| • Abilities of technological design | X | X | X |
| • Understandings about science and technology | X | X | X |
| • Abilities to distinguish between natural objects and objects made by humans | X | | |

Source: National Research Council. (1996) *National Science Education Standards*. Washington, DC: National Academy Press.

In postsecondary science courses, technological advances have resulted in curricular changes (e.g., Kincannon 1994; Jungck et al. 1997). Entire curricular strands, which focus on the interplay of science, technology, and society (STS), have been developed; schools of engineering have often taken the lead in developing such strands. In some undergraduate science courses, technology and its enabling by science is covered. For example, courses in life sciences may refer to genetic engineering techniques, and biochemistry may include gene sequencing (Heady 1994). In interdisciplinary college-level introductory science courses, the science behind health and environmental concerns arising from advances in technology may be covered. The point is that in the future our students must be prepared to live with technology that will prompt them to ask and to answer questions about nature, questions that today we cannot even imagine. Science and technology will take them there.

# From the Field

## Issues in Science, Technology, and Society

Gordon Johnson, Department of Physics
University of Northern Arizona

Learning science as you encounter it is the essence of scientific literacy. Citizens need to be able to make informed decisions on issues that have a scientific and/or technological connection. Yet, very few people have the luxury of preparing themselves to make informed decisions on these issues by studying adequately the traditional sequences of discipline-based science and technology. Thus, very few people will be able to maintain a functional level of scientific literacy.

Opportunities for students to demonstrate to themselves that they can learn sufficient science for intelligent decision-making are needed. Issues in Science, Technology, and Society is a course designed to provide such an opportunity. Students are challenged to gain science understandings in the context of societal issues that are of interest to them. Specific science concepts are learned because they have a direct connection to a real concern. Through a variety of issues, students encounter the characteristic features of science and technology and the differences between science and technology. Students learn science because they come to see the potential contributions of science and technology to the resolution of societal problems.

In addition to providing a context for learning science, the course encourages student choices. Through a brainstorming activity, students identify current issues of significance and interest to them. They then, with instructor guidance, develop a list of criteria for choosing issues. Based on those criteria, students vote for their preference of topics to investigate. The three topics of highest interest are chosen for class investigation. Small groups of individuals with interests in a common topic are also identified. Those small groups carry out an investigation of an issue related to that topic for both written and oral presentations. Each individual also selects a specific topic of interest for an independent investigation that culminates in a written report.

After the initial identification and choice of topics to study, the topics are further refined and stated as issues. For example, the topic "global warming" was focused to "Should the U.S. Senate ratify the Kyoto Protocol?" Students develop relevant subquestions in their small groups. Being able to generate appropriate questions is an important skill. The questions are collected and categorized by the instructor (without any editing) as science-, technology-, or society-based. In later units, students categorize the questions. For class investigation of the initial issue, the instructor models the approach to be used for the small groups and individual investigations. In the class investigations, the students gradually take more and more responsibility for researching the subquestions. A specific set of procedures, with some flexibility, is followed in all of the investigations.

Current articles from journals such as *Scientific American, Environment, World Watch, National Geographic, Technology Review, New Scientist,* and *Smithsonian,* along with numerous Web sites, are used as references for the course. The latest "State of the World" published by the World Watch Institute is also a frequently used reference. The instructor chooses specific articles from these resources for required background reading. Part of the evaluation for each issue investigated by the class is a reaction paper. Based on what students have gained from the investigation, they react in written form to a current article (chosen by the instructor) that is related to the issue. The reaction paper gives a good indication of the level of understanding of the student with respect to that issue and of his or her ability to apply that understanding.

Students selected a broad range of issues for investigation during a recent semester. In addition to the one mentioned earlier, "Should the U.S. Senate ratify the Kyoto Protocol?", the issues related to the current supply of material resources, animal organ transplants, reintroduction of the Mexican grey wolf, multiple-birth technology, the Y2K problem, Glen Canyon Dam,

growth plans for the local community, biological and chemical weapons, alternative energy sources, population control, use of a waste tax, and labeling of genetically engineered foods. The small groups used a variety of approaches in their presentations. I especially encouraged presentations that would elicit whole-class discussion.

Assessment of student achievement included the reaction papers, abstracts of required readings, student response to small-group presentations, written papers on individual investigations, presentation and participation in small-group investigations, and performance on two essay examinations. The examination questions called on students to evaluate critical issues and to defend particular positions with respect to those issues.

Student response to the course has been very positive. The students appreciated the chance to formulate and redefine their own thoughts with respect to issues, to hear and respond to the thoughts of others, and to be introduced to sources of information they normally would ignore. Attitudes and positions on issues were surveyed prior to and following the investigations. Clearly, positions and attitudes changed as students were exposed to more information and the separate but interdependent roles played by science and technology become more apparent. Students also came to realize the responsibility that society has to direct technology and to contribute to decision-making about technological issues.

The Issues in Science, Technology, and Society class is made up of students from a variety of fields, including the arts, science, history, English, business, and forestry. A majority of students, however, plan a career in science teaching. The prospective science teachers especially benefit from the interaction across discipline boundaries; they develop the ability to communicate science in a nonthreatening way to peers with different backgrounds. And all class members come to appreciate the importance of considering a variety of factors in making decisions. The overall goal of improved scientific literacy is the focus of the course, but an important component—the better understanding and appreciation of the endeavors of the scientific and technological communities—is also realized.

# Science in Personal and Social Perspectives

STANDARD
F

Content Standard F emphasizes that science is a part of everyday life. An understanding of science adds to the ability of citizens to make good decisions at the personal and societal (and political) level. When we personalize science, it becomes obvious that science is not value neutral—that there is "good" science and "bad" science. The Standard notes the importance of understanding the interrelationship of humans and science, particularly in regard to health and the environment at the personal, community, and global level.

## Table 4.5

*Content Standard F: Science in Personal and Social Perspectives—Understandings That Students Should Develop in K–12 Activities*

| Grade Level | | |
| --- | --- | --- |
| K–4 | 5–8 | 9–12 |
| • Personal health | • Personal health | • Personal and community health |
| • Characteristics and changes in populations | • Populations, resources, and environments | • Population growth |
| • Types of resources | • Natural hazards | • Natural resources |
| • Changes in environments | • Risks and benefits | • Environmental quality |
| | | • Natural and human-induced hazards |
| • Science and technology in local challenges | • Science and technology in society | • Science and technology in local, national, and global challenges |

*Source:* National Research Council. (1996) *National Science Education Standards.* Washington, DC: National Academy Press.

The relation between Content Standard E (Science and Technology) and Standard F (Science in Personal and Social Perspectives) cannot be ignored. Science and technology shape our world—and pervade our lives. We need to see these areas with a social and personal perspective in order to live productively and with an appreciation for our surroundings.

## From the Field

### Value-Based Science—Chemistry and the Environment

Theodore D. Goldfarb, Department of Chemistry
State University of New York at Stony Brook

For the past twenty years I have taught an undergraduate environmental chemistry course. It is designed to satisfy the university's general education requirement on the social and global implications of science, and requires only one college-level chemistry course as a prerequisite. Thus, the course attracts many non-chemistry majors, including a minority of humanities and social science students.

The course content and emphasis reflect my firm belief that an understanding of the chemical aspects of environmental issues requires an examination of the technical aspects of those issues within a broad philosophical, sociopolitical, and economic context. If students are taught about only the scientific and technical dimensions of environmental problems, they will develop the false impression that sustainable solutions often can be achieved by relatively simple technological fixes. They will fail to appreciate that a process that looks good within the narrow confines of a scientific laboratory may not translate into a useful technology in the much more complex and demanding conditions of the real world.

An environmental course is an ideal venue for emphasizing the fact that science is a human endeavor. The investigation of the scientific dimensions of virtually all environmental issues will reveal that rather than being objective and value-free, they are infused with a wide variety of ethical considerations related to the need to make value-laden choices. When my students are introduced to risk-benefit analysis, which is frequently prescribed as the scientific approach to making decisions about proposed solutions to environmental problems, they quickly learn that such analysis is far from being a straightforward, objective procedure. Comparing risks and benefits invariably requires devising a common scale by which to evaluate very different phenomena. This task reflects the values of the individuals performing the analysis. In many cases, the benefits of the action under consideration may not accrue to precisely the same group of individuals who must endure the risks. Under such circumstances, a decision must be made as to whether this fact renders the analysis as ethically unacceptable. Even the initial task of deciding on the list of risks and benefits that are appropriate to include in the analysis will usually require decisions that are not clear-cut and may lead to controversy among stakeholders with different value-determined perspectives. Once the list has been determined, the task becomes to examine the scientific evidence to evaluate the magnitude of each risk and benefit. Rarely is this evidence sufficient to make such assessments without invoking a variety of assumptions that reflect the judgment and therefore the values of the assessor.

Each year I include the Love Canal (Niagara Falls, New York) controversy as an instructive issue that reveals many of the common social complexities of an environmental situation with significant public health implications. We begin with a careful examination of what was known in 1978 about the toxic chemicals that had been dumped in the Love Canal site, the exposure of nearby residents to these toxins, and the results of surveys that had been performed on the health of the residents. We then read about the conclusions reached on the basis of these facts by spokespeople for the Love Canal Homeowners' Association, volunteer scientists who served as advisors to this group, other independent scientists, the New York State Commissioner of Health, and state and federal government scientists with regard to what remedial actions should be taken and what additional scientific and health studies needed to be done. A classroom discussion of the sharply differing recommendations made by these individuals invariably leads students to conclude that these different interpretations of the facts have more to do with values than with science. Furthermore, the motivating value differences are understandable primarily in terms of the controversy.

A study of the consequences of human exposure to asbestos provides another opportunity to examine the influence of values on scientific research. Students consider the fact that almost all of the scientific studies done by scientists employed by the asbestos industry or its insurers prior to the 1960s exonerated this widely used material from any significant causal role in human disease. In sharp contrast to these results, subsequent studies done by independent scientists revealed that asbestos is a potent carcinogen and a cause of other seriously debilitating pulmonary disorders. An examination of the studies reveals that studies that were paid for by the industry had a common flaw. They excluded from the study group those with the long-term exposure necessary for asbestos to induce cancer and lung disease. This raises the question of whether those who designed these faulty studies were guilty of intentional fraud. Most students conclude that this is not very probable. They argue that the poorly designed studies were more likely to have been a result of carelessness, which was subtly, but not consciously, related to the researcher's self-interest.

Teaching such a course poses an important ethical issue for the instructor. Is it necessary or appropriate to try to maintain a strictly objective posture and not reveal one's own conclusions when teaching about controversial issues? I have come to the conclusion that this is virtually impossible. When I tried to do this in the past, I found that students invariably picked up on clues about where I stand. Furthermore, feigning objectivity and refusing to respond to inevitable direct inquir-

ies about my views set a bad example in a course where my goal is to get the students to develop opinions they can defend and reach conclusions about the controversies they study. It is, of course, necessary to be as fair as possible in exposing students to all of the conflicting arguments and evidence in a controversy. Students should be made to feel comfortable about challenging the opinions of the instructor. Even more difficult is the ethical imperative of basing a student's grade on the quality of his or her work although it may be in support of a conclusion with which you strongly disagree!

## History and Nature of Science

At some points in the past—and this is particularly true in the media—scientists have been portrayed as some sort of gods who possess knowledge that can be known only to them. Content Standard G emphasizes instead that science is a human endeavor. And because science is carried out by humans, it is a subject that anyone may pursue. To that end, the philosophy of the *National Science Education Standards* is inclusive; the *Standards* continually emphasize that science should be accessible to *all* students.

## Table 4.6

*Content Standard G: History and Nature of Science—Understandings That Students Should Develop in K–12 Activities*

| Grade Level | | |
| --- | --- | --- |
| K–4 | 5–8 | 9–12 |
| • Science as a human endeavor | • Science as a human endeavor | • Science as a human endeavor |
| | • Nature of science | • Nature of scientific knowledge |
| | • History of science | • Historical perspectives |

*Source:* National Research Council. (1996) *National Science Education Standards.* Washington, DC: National Academy Press.

The concept that there is a history of science is often neglected in science courses. Occasionally textbooks will point to isolated incidents such as the experiments that led to the nuclear model of the atom or the origin of quantum mechanics. But generally such developments are left to history departments, which may offer a History of Science course, and philosophy departments, which may offer a Philosophy of Science course. A sense of history can give insight as to why certain questions were being investigated by scientists and why some answers, which may seem obvious now, were not apparent in earlier times. To understand why science has developed as it has, students need to understand the cultural contexts in which scientists found themselves.

There are other reasons for addressing the history and nature of science in science courses. In science teaching, blending the historical and social elements behind the research may "bring to life" the work for the student and organize for him or her a large and disjunctive body of work. Rice (1994) argues that putting science in a humanistic context shows students about different research approaches and values of scientists.

In addition, history can portray the role of "ordinary" and diverse people who have made connections and advanced science. It provides yet another way to open the doors of science to persons who are representative of the community at large. At the very least, it makes science more acceptable to a public that is increasingly responsible for making individual and societal decisions that *should* be based on science. The following vignette illustrates how these concerns led to the development of Project Inclusion, a project that provides historical perspectives of chemistry from diverse cultures.

## From the Field

### Project Inclusion: Using the History of Diverse Cultures to Facilitate the Teaching of Chemistry

Janan M. Hayes, Science Division, Merced College
Patricia L. Perez, Chemistry Department, Mt. San Antonio College

About seven years ago, we had an intense discussion about our current teaching methods, materials, and students compared with those from our initial years in the classroom. (We have been collaborators on various projects since the early 1970s.) Intuitively, we have always believed, as do many science educators, that our discipline is universal, that chemistry is practiced everywhere and anywhere by *all* persons in some fashion with no regard for geographical boundaries, cultural limits, or time constraints. And we believe that this universality must be conveyed to all students in every chemistry course, whether it be at the introductory, intermediate, or advanced level. The major change in the undergraduate classroom over the years has been in the faces of the students. We are now dealing with a much more diverse population (in terms of gender, ethnicity, and skills) than we did in our early classes, which were fairly homogenous groups, and our concern is that our courses convey the relevance of chemistry and science to students of all backgrounds.

A serendipitous happening almost seven years ago was a catalyst toward a solution for our concern. Jan happened to read an article in *American Antiquity* (Hewitt, Winter, and Peterson 1987) that theorized that the indigenous people in pre-Hispanic Oaxaca, Mexico, obtained valuable salt by a series of solar evaporations of brine from a spring, as did many peoples around the world. The next day she told this story in a beginning chemistry class while introducing the topic of solutions and concentrations. It caught the attention of five Mexican women students. These pre-nursing students had made limited progress in the course and were considering dropping it. The story of the ancient Oaxacans' work was the "hook" the students needed to make a strong connection to the chemistry course and successfully complete it. A short time later, in reviewing general chemistry textbooks for possible adoption, Pat focused on their "diversity content." She noted the paucity of pictures of women and minority chem-ists. In addition, the examples used to illustrate the history of chemistry and the topic under discussion were taken primarily from American and European chemistry.

An idea began to develop that we could better reach our students if we had course materials that related chemistry and physical science to our students' cultural backgrounds and to the historical development of science and technology in their countries of origin. Our work has evolved into Project Inclusion (PI). For the past seven years, we have been researching, writing, and disseminating resource materials that highlight the chemical contributions of women and members of other under-represented groups over a wide range of cultures and times. These resource items (we call them bits, bytes, and modules) may vary in length from a few minutes of class time to those that take a class period (C&EN 1996; Hayes and Perez 1997). We have used these materials in many different ways—for example, a few minutes in lecture with an example to illustrate the topic under study, ten to twenty minutes on the life and work of a scientist as part of a laboratory lecture, forty-five minutes for a quick trip around the world of chemistry in a course prologue, and as the basis of a short research paper by a team of students for extra credit. Each item provides a resource that can help our students understand that chemistry is not only universal but also historically relevant and specifically a part of their own culture, whatever that may be.

Materials have been disseminated to teachers at all grade levels, including those in colleges and universities, through summer Project Inclusion workshops, which have been held in northern and southern California as well as in Pennsylvania. Activities at the workshops include examples of ways to use existing materials, the opportunity to create new materials, and a field trip to a local indigenous chemistry site. For example, Pat met Julia Parker, a Native American basket maker and storyteller, at the February 1998 meeting of the California Indian Conference. A subsequent visit with Julia at the Indian Museum in Yosemite Village led to a field trip to Yosemite National Park as part of the summer 1998 Project Inclusion faculty workshop. Julia shared the details of the traditional methods she uses to gather, prepare, and dye the materials that end up as black patterns in her beautiful baskets. Then the workshop participants investigated the science—a little redox chemistry of iron (since the major ingredients in the dye are elemental iron, ferric oxide, and a plant tannic acid source). At the summer 1999 PI workshop, we discussed the traditional methods used by the Gabrieleno Tongva people (of Los Angeles and Orange Counties in California) to produce dyed baskets. Again, it appears that similar iron redox reactions are the basis of the dye color. What a wonderful way to help students see that the redox equations we are balancing in class have real world applications.

Pat has had Egyptian students open the world of Islamic alchemy to her and their fellow students in a short, extra-credit research paper because they wanted to learn of "their" chemistry. One of Jan's students did an Internet research project on Kakiemon Red, a red dye used in seventeenth-century Japanese pottery making, as described in a Nippon Steel ad in *Smithsonian* magazine. The pottery is still made using traditional methods, but the science of the color process did not become clear until the 1950s when Toshio Takada began to study the ferro-magnetic structure of the colorant. The work is being continued by his son, Jun Takada. Their scientific studies show that particle thickness of the glaze and temperatures of the various steps of the process must be carefully controlled to produce Kakiemon Red, which is alpha-iron(III) oxide. This becomes significant to our students when we point out that Toshio Takada also has the patents on gamma-iron(III) oxide, which is the magnetic material in recording tapes. Such wonderful iron chemistry—and most of the PI investigative work on the topic—was done by a future elementary school teacher who has had only one introductory chemistry class!

We all recognize that our student populations today are more diverse than those of a few years ago; the authors know that these resource materials from history, various cultures, and many locations have helped students succeed in their own science classes. Look at the names and faces of your students; review your local resources.

You will probably be able to find ways, as we have, to use the history of diverse peoples as a catalyst to forge a bond between the world of the student and the world of science.

## References

Bronowski, J. (1990) *Science and Human Values*. New York: HarperCollins.

C&EN (1996) Modules Aim to Add Multicultural Dimension to College Chemistry. *Chemical and Engineering News*, April 29, 70.

Hardin, G. (1968) The Tragedy of the Commons. *Science* 162, 1243-48.

Hayes, J., and Perez, P. (1997) Project Inclusion: Native American Plant Dyes. *Chemical Heritage* 15 (1), 38-40.

Heady, J. (1994) Ethics Discussions about Technologies Such as Those Associated with In Vitro Fertilization, the Genome Project, and Gene Replacement. In *Science Discoveries and Science Teaching: The Link,* edited by E. Siebert and C. Estee. Cedar City, UT: Society for College Science Teachers.

Hewitt, W. P., Winter, M., and Peterson, D. A. (1987) Salt Production at Hierve El Agua, Oaxaca. *American Antiquity* 52 (4).

Jungck, J. R., Soderberg, P. Stanley, E., and Vaughan, V. (1997) Computer-Enhanced Collaborative Learning. In *Methods of Effective Teaching and Course Management for University and College Science Teachers,* edited by E. Siebert, M. W. Caprio, and C. Lyda. Dubuque, IA: Kendall/Hunt.

Kincannon, E. (1994) Chaos in the Undergraduate Curriculum. In *Science Discoveries and Science Teaching: The Link,* edited by E. Siebert and C. Estee. Cedar City, UT: Southern Utah University, Society for College Science Teachers.

Lumsden, A. S. (1997) The Large Class. In *Methods of Effective Teaching and Course Management for University and College Science Teachers,* edited by E. Siebert, M. W. Caprio, and C. Lyda. Dubuque, IA: Kendall/Hunt.

National Research Council (NRC). (1996) *National Science Education Standards*. Washington, DC: National Academy Press.

Rice, R. (1994) The Structure of DNA and the Humanistic Study of Science. In *Science Discoveries and Science Teaching: The Link,* edited by E. Siebert and C. Estee. Cedar City, UT: Society for College Science Teachers.

# Science Education Program Standards

| Eleanor D. Siebert |
|---|

**T**he Program Standards of the *National Science Education Standards* invite us to look beyond the individual courses we teach. They ask us to consider the sequence of courses and educational experiences—that is, the program of study—in which our students are involved. At the university and college level, undergraduate students engage in a multiyear program of study that leads to a degree at the baccalaureate or associate level. The program generally focuses students in an area of study by designated course work and experiences that lead to an academic major, which will be in one or more of the science departments or divisions. Although written for K–12 science programs, the six Program Standards provide a guide for assessment of science programs at the undergraduate level. They address program consistency, ensuring that all elements are directed toward the same goals. The Program Standards also state that studies of science should be connected to other academic areas—*especially* mathematics—and that adequate resources and equal opportunities exist for all students. These Program Standards are:

**A** All elements of the K–12 science program must be consistent with the other *National Science Education Standards* and with one another and developed within and across grade levels to meet a clearly stated set of goals.

**B** The program of study in science for all students should be developmentally appropriate, interesting, and relevant to students' lives; emphasize student understanding through inquiry; and be connected with other school subjects.

**C** The science program should be coordinated with the mathematics program to enhance student use and understanding of mathematics in the study of science and to improve student understanding of mathematics.

**D** The K–12 science program must give students access to appropriate and sufficient resources, including quality teachers, time, materials and equipment, adequate and safe space, and the community.

**E** All students in the K–12 science program must have equitable access to opportunities to achieve the *National Science Education Standards.*

**F** Schools must work as communities that encourage, support, and sustain teachers as they implement an effective science program.

*Source:* National Research Council. (1996) *National Science Education Standards.* Washington, DC: National Academy Press, 210–224.

Because the lessons taught by science are central to many aspects of life, science courses are required by many postsecondary programs; science departments heavily "service" some programs, such as pre-professional health, and provide less service to other programs, such as education. In most universities, students in all undergraduate programs must take at least one science course that will deepen their understanding of and appreciation of science—that is, as "educated individuals" they are expected to be scientifically literate. As a result of the many audiences that science courses serve, departments of science and faculty who teach courses for nonmajors have a special obligation to understand the objectives of programs beyond those in our majors and the role of science in those programs. It is important to realize that the students in our classes may be following a course of study much different from our own; they will see science and its significance from a much different perspective than we do. These students will attach value to science as it relates to their lives and, often, as it relates to their field of study; the standards indicate that we must be concerned with these students. However, the focus of this chapter on program standards is on undergraduate science programs, which are generally under the jurisdiction of the academic department or division of which we as individuals are a part.

## Program Consistency

All elements of a K–12 science program must be consistent with the other *National Science Education Standards* and with one another and developed within and across levels to meet a clearly defined set of goals.

- In an effective science program, a set of clear goals and expectations for students must be used to guide the design, implementation, and assessment of all elements of the science program.

- Curriculum frameworks should be used to guide the selection and development of units and courses of study.

- Teaching practices need to be consistent with the goals and curriculum frameworks.

- Assessment policies and practices should be aligned with the goals, student expectations, and curriculum frameworks.

- Support systems and formal and informal expectations of teachers must be aligned with the goals, student expectations, and curriculum frameworks.
- Responsibility needs to be clearly defined for determining, supporting, maintaining, and upgrading all elements of the science program.

*Source:* National Research Council. (1996) *National Science Education Standards.* Washington, DC: National Academy Press, 210.

Program Standard A indicates that an effective program of science study is consistent with a clear set of goals and student outcomes. At the undergraduate level, the key to designing such a program is to begin by reaching faculty consensus in identifying goals and student outcomes. The shared and common vision about what ought to unite a science program establishes a framework for the curriculum, which can then be developed as a sequence of courses and experiences that will provide a route to achieve those goals. However, curriculum development alone is not sufficient to ensure adequate opportunity for student learning. The linking of curriculum to student outcomes and the development of effective pedagogical approaches are essential to produce well-educated science majors (see Chapter 1). Program Standard A also recognizes that programs should continually evolve. Some factors influencing programmatic change include increasing science knowledge, shifting economic bases, emerging technology, and changing student experiences. Program assessment—which is almost continual, variable in form, and aligned with program goals—should be used to guide this evolution. In other words, program assessment is integral to the good health of any department—and to any educational institution.

# From the Field

## Program Reviews: A Vital Component for Quality Undergraduate Science Education

Eleanor D. Siebert, Department of Physical Sciences and Mathematics
Mount St. Mary's College

Mount St. Mary's College in Los Angeles offers a liberal arts, values-oriented education to a diverse student body. Embedded in its mission statement is a commitment to programs devoted to "excellence in the liberal arts and sciences and career preparation at the associate, baccalaureate, and master's degree levels, with a special focus on education of women for participation and leadership in our society and our times." The student body reflects the ethnic diversity of California and Los Angeles, having no single racial or ethnic group that currently makes up a majority.

At Mount St. Mary's, all academic and co-curricular programs are on a five-year review cycle. The purpose of the review is to ensure that program outcomes are consistent with the mission of the institution, its short-term strategic plan, student needs, and professional expectations. The review initiates a reflective self-study that enables the program faculty and staff to identify and clarify its strengths and weaknesses and to more accurately focus its energies to meet the academic needs and concerns of the students and the institution. Reviews in which faculty are most intimately involved are academic reviews on either the departmental or divisional level.

The academic review focuses on seven major areas: program goals; curriculum/experiences; faculty and staff; student services; student outcomes; contribution to the college community; and responses to previous reviews. An external committee, composed of faculty from many departments, oversees the review and offers a college-wide perspective to guide the process.

The first step in the self-study is an examination of program goals—the beacons that keep a program on track and help in making decisions between reviews. In the past, many programs expressed these goals in terms of a departmental mission or purpose; however, more recent reviews at Mount St. Mary's College have put students at the center of the process and consider goals in terms of student outcomes specific to the program. While there is no single way to carry out a program review, the first step is virtually always an examination of program goals and

achieving a consensus of vision among the faculty and administrators involved with the program.

After department members set goals and articulate student outcomes, the second and perhaps more important step is to outline strategies for achieving these goals and outcomes. Strategies require an examination of the *curriculum*, of *faculty members* (and the ways in which they contribute to reaching the goals of the program), and of *support services*—especially academic advising—offered to students taking courses in the program. In the past, developing the strategic plan has taken much time, requiring many hours of face-to-face discussions, reflections, and debate. The process is facilitated when all department faculty are involved in developing the program goals and student outcomes and when the final version is reached by consensus.

The third step in the review process involves determining progress toward program goals by the measurement of student outcomes. Information collected by the department and used to assess student outcomes includes data on program graduates, academic performance in upper-division courses, retention and recruitment of students in the major program, and feedback on students who participate in summer programs at other institutions. At Mount St. Mary's College, the Office of Institutional Research routinely gathers information useful to the review. Such data include grade distributions, course evaluations, departmental cost analyses, standardized test scores, and student surveys. As important as the assessment information is, however, it is worth nothing if not used properly in assessing how well a program achieves its goals.

The program is then viewed by departmental faculty in the context of the institution's goals. This view goes beyond the program offered to majors within the discipline or a related discipline and considers contributions to the liberal arts program at the college. At Mount St. Mary's College, there are seventeen student outcomes that are considered to constitute the essence of a liberal arts education—an education that provides students the opportunity for "knowledge and appreciation of the diverse fields of human endeavor" (*Mount St. Mary's* 1998–2000). Any departmental courses taught for "general studies" credit must address several of these components. Of the seventeen components, those particularly relevant to the sciences include:

- Effective written expression of ideas;
- Effective oral communication;
- Analysis of assumptions, methods of argumentation, values;
- Problem solving: defining problems, identifying issues; organizing, analyzing, synthesizing ideas; comparing/contrasting ideas; decision making;
- Curiosity about and a spirit for investigating the natural universe;
- Ability to recognize patterns of thought used in science and mathematics;

- Understanding of the impact of advancing technology on human society and culture; and
- Understanding of criteria and standards to assess personal moral values and ethical judgments.

The department examines each of the courses that it offers for General Studies credit to determine how the course addresses the outcomes and how the outcomes are measured. The final version of the departmental or program self-study is forwarded to the collegewide Curriculum Committee, which has faculty representation from across the many programs at the college. The committee, charged with the review of all college programs, makes recommendations to conclude the process.

All planning and discussion—at both the departmental and collegewide levels—are carried on in the context of previous reviews. The department examines its responses to previous concerns and subsequent development of its strengths. The review takes the better part of a semester to complete at the department level, but the results are a well-marked road toward visionary goals—of the program and the institution. Programs with goals supported by well-defined student expectations and well-designed curriculum frameworks provide students with opportunities for seamless intellectual growth. That is certainly worth a semester's work!

# Curriculum Criteria

The program of study in science for all students should be developmentally appropriate, interesting, and relevant to students' lives; emphasize student understanding through inquiry; and be connected with other school subjects.

● The program of study should include all of the content standards.

● Science content must be embedded in a variety of curriculum patterns that are developmentally appropriate, interesting, and relevant to students' lives.

● The program of study must emphasize student understanding through inquiry.

● The program of study in science should connect to other school subjects.

*Source:* National Research Council. (1996). *National Science Education Standards.* Washington, DC: National Academy Press, 212.

To be aligned with the *National Science Education Standards*, a postsecondary science program—and every course—should be inquiry-based and connected to students' lives and to other courses within and without the sciences. Moreover, courses should be carefully sequenced in order to allow students to construct successively more complex understandings on a firm knowledge base. That base will include information from across the sciences and may involve critical thinking and writing courses. The following vignette illustrates how conservation science provides the central focus for many programs at the College of Santa Fe and how the principles of Program Standard B were applied in curriculum development.

## From the Field

### Science for All Students: Taking This Mandate Seriously
Robert W. Harrill, Executive Director
Institute for Conservation Studies
The College of Santa Fe

Reflecting on forty years of experience in science and education as a student, teacher, researcher, administrator, program developer, activist, and employer, I marvel at the lethargy of the higher educational system in responding to the frequent and often re- peated call to make science relevant to all students' lives. Having just returned to higher education after a decade-long absence working in the "real world" of applied conservation science, I am troubled that because of our inability to create enough excitement and interest about, and relevance in, our science programs we must continue to *require* students to take our courses (at least one or two). Do we really believe this will produce a scientifically literate citizenry? Experience tells me that the vast majority of students who are forced to take our courses for nonscience majors do as little as possible to satisfy the science requirement and move on with enthusiasm to the courses in other departments that really interest them.

The challenge is to develop programs that will attract the majority of students in our institutions (not just the majors in traditional science disciplines), who will not only meet the minimum requirements we establish but will also choose to take more courses than we require. Responding to this challenge means that we must either modify or expand the purposes of traditional science departments or create new programs outside the traditional departments. The former seems unlikely since the primary purpose of traditional departments has been and continues to be the training and education of the next generation of specialists—those majoring in specific science disciplines. The "service func- tions" of traditional departments represent a relatively small investment of resources in most colleges and universities. Yet if we are to take the principles reflected in the *National Science Education Standards* seriously and apply them to higher education, we must raise the importance of science education for the majority of college students and create programs that will captivate and hold their inter-

est beyond the bare minimum required for graduation. In other words, we must create programs in which "nonscience" students are integral participants.

A prototype of such a program has been developed recently through the Institute for Conservation Studies at the College of Santa Fe (Harrill 2000). We have designed and initiated a project-based approach to undergraduate education whereby students of the natural and social sciences, arts, and humanities come together as members of project teams to work side-by-side with professionals on active conservation projects. The ICONS program (as it is called at the college) has multiple purposes. It is a program that:

- integrates academic knowledge, field-based assessments, and new technologies for landscape analysis by (1) applying field assessment techniques in conservation projects to document existing ecological conditions and monitor the effects of land use practices, and (2) applying remote sensing and geographic information system technologies to help resolve conflicts over land and resource use;

- works to reverse the trend toward impoverishment of natural communities through applied research on effective ways to restore proper functioning of native ecosystems and their inhabitants;

- offers experience and training to show how people can learn to live within an ecosystem in a way that sustains the social and ecological health of their place; and

- expands awareness and understanding of conservation issues, problems, and solutions for a broad audience through creative use of words and images.

With a foundation in conservation science, ICONS is a hands-on, experiential program that engages students as active participants in a learning process balancing conservation theory and practice. Three aspects of the program appear to offer special appeal to our students, few of whom come to the college to study science:

- First, by participating in applied field projects, students feel that they are doing something real, meaningful, and useful, even when the level of participation is superficial (as it might be in the freshman year). This experience is a powerful motivator for future learning. As additional knowledge and skills are acquired, students feel increased confidence and a strong sense of making a contribution.

- Second, the issues we address in our projects are interesting and important to a diverse student audience. In fact, we are finding that students who previously avoided science at all costs are joining the program because of their interest in "wildlife" or "protecting nature" and are rediscovering their interest in learning science.

- Third, conservation work is sufficiently diverse that students from virtually every academic department can find a place on a project

team and contribute to the work being done. Recent examples include a studio art major who created an interpretive poster illustrating the wild mushrooms found in the Alto Merse Nature Reserve in Italy and a group of moving image arts majors who are created a thirty-minute documentary video for the Sustainable Uses of Biological Resources Project in Ecuador. The soundtrack for that video, created by a music major, includes sound recordings of community ceremonies in the Afro-Ecuadoran village of Playa del Oro as well as natural patterns of sound from the surrounding forests and rivers. The most surprising result of the ICONS program to date is the extent to which nonscience students want to participate in our conservation projects.

Conservation projects in which ICONS students have participated include making riparian and forest health assessments in the Santa Fe River watershed, assessing the impact of recreational activities on fragile ecosystems in Canyonlands National Park, developing visitor information guides (natural history) in Bandelier National Monument, biological mapping in the Alto Merse Nature Reserve (Tuscany), measuring the impact of elk on native vegetation communities (Bandelier National Monument), inventory and assessment of potential old growth forest stands (Santa Fe National Forest), documenting the Sustainable Uses of Biological Resources Project (Ecuador), and assisting Mayan communities to develop a management plan for the Sarstoon Temash protected area (Belize).

The ICONS project model promotes the integration of learning from diverse disciplines. Project teams, which may be composed of students whose primary interests range from conservation science to film, education to photography, work together on a common problem. Each student brings knowledge, skills, values, and experience to the process, while learning from the others. The process of interacting, sharing, debating, resolving conflicts, and working toward a common goal is a key component of the project model. Student interest in the program has stimulated dialogue with faculty from different departments and has generated strong interest in interdepartmental collaboration. We now have secured institutional approval for two majors: a B.S. in Conservation Science and a B.A. in Conservation Studies with interdepartmental concentrations in conservation policy and planning, the conservation documentary (film, video, and photography), conservation writing, and conservation education. Applied conservation projects become the integrating element for all of these concentrations and encourage active collaboration among faculty from different specialties as we attempt to meet the needs of growing numbers of students participating in the ICONS program.

ICONS is a new program with a limited history: two years in development and seven semesters of operation (in-cluding two summer sessions). Its short existence precludes a comprehensive evaluation of its impact on science and nonscience students. Preliminary trends are encouraging, however. Student participation is increasing dramatically, with about one-third science students and two-thirds students from other departments. Nonscience students are choosing to take advanced courses in conservation science that satisfy no requirement and to participate in ICONS projects because they want to. Faculty from nonscience departments are calling us to find out what is going on with ICONS and to see how they can participate. Project opportunities are developing faster than we can find participants. At its current rate of growth, we expect over one-third of enrolled students (over 275 students) to participate in one or more offerings in the ICONS program within the next year or two.

A project-based program does require changes in the way the institution operates. The college needs to dedicate funds for developing and running field projects and to support faculty whose primary function is coordinating and operating field projects. We have chosen to design the program from the ground up with all course offerings developed specifically to meet the needs of the program. We offer a multilevel approach that allows science and nonscience students to participate in the same courses. The student who wants to make conservation documentary films needs a different level of understanding of conservation science than the student who wants to do research on ecosystem restoration. Perhaps the most significant factor contributing to the success of the project-based approach is formal recognition by the institution of the validity of a program that is not based in a single department (though conservation science is the foundation on which the ICONS program is built) but is a program that attempts to engage students and faculty from all departments in an integrative educational effort.

## Science and Mathematics

STANDARD C

The science program should be coordinated with the mathematics program to enhance student use and understanding of mathematics in the study of science and to improve student understanding of mathematics.

*Source:* National Research Council. (1966) *National Science Education Standards.* Washington, DC: National Academy Press, 214.

For those of us teaching college and university introductory science courses—especially in the physical sciences—Program Standard C should be welcome news. Many students find that the greatest barrier to being successful in science at the introductory level is their lack of facility with mathematics and mathematical reasoning. Perhaps the most obvious aspect of the relation of science and mathematics is through the use of numbers. Scientific data frequently involve numerical measurements, and additional numerical manipulations are often involved in order to relate the raw measurements to hypotheses. The interpretation of these numbers and an understanding of their significance may involve statistics.

But the connection between science and mathematics goes beyond the mere use of numbers. Mathematics involves logic, and its study is characterized by doing and reflection—just as is required in the study of science. Moreover, advanced theoretical understanding of *all* sciences has mathematical underpinnings—this is true in quantum chemistry, physics, modeling of biological systems, and the Earth and space sciences. Thus, the coordination of science programs with mathematics programs at the university/college level may be even more important to our science majors than our students would like to admit. It is reassuring to know that *Standards*-based teaching in K–12 will reinforce this connection throughout the college preparatory years; but it is a connection that should not be lost at the undergraduate level. The following vignette describes how a consortium of colleges and universities focuses on improving learning by emphasizing connections among mathematically based disciplines.

# From the Field

## Linking Mathematics and Science: The Long Island Consortium for Interconnected Learning[1]

Jack Winn, Department of Mathematics
State University of New York–Farmingdale

The Long Island Consortium for Interconnected Learning in Quantitative Disciplines (LICIL), directed by Alan Tucker (State University of New York [SUNY]–Stony Brook), is a consortium of faculty at ten colleges and universities on Long Island. The consortium is a four-year project funded by the National Science Foundation. A fundamental premise of LICIL is that students learn better when the natural connections among the mathematically based disciplines are emphasized. In particular, many of the LICIL projects are directly based on the link between mathematics and the natural sciences. LICIL activities on the participating campuses have resulted in a substantial increase in interdisciplinary cooperation, curriculum reform, and instructional innovation.

LICIL faculty projects have encompassed such general themes as educational technology; multidisciplinary courses; precalculus, calculus, and postcalculus reform; alternatives to lecturing such as collaborative learning; interdisciplinary student workshops; interdisciplinary student projects; mathematical modeling in the real world; interactive teaching; learning communities; and computer-generated and computer-graded personalized homework. Examples of particular projects carried out on the campuses are Using Computers to Analyze Data in Biology Labs; Use of the Graphing Calculator and CBL in the Chemistry Lab; Mathematical Modeling Workshops for General Chemistry; Mathematical Thinking and Earth Science Concepts; Learning Modules to Illustrate Linkages Between Periodic Functions Taught in Mathematics and Physics; Integrated Mathematics-Physical Science Learning Labs; Levodopa to the Brain–A Rate/Change Study; and Interdisciplinary Workshops in Mathematics and Physics—to give only a very small sample of the many interdisciplinary LICIL activities.

At one consortium school, SUNY–Farmingdale, the chemistry department redesigned its laboratories around the graphing calculator as a direct consequence of the mathematics department's adoption of the TI-85 graphing calculator (now the TI-86) in

---

[1] The LICIL Web site is *www.LICIL.org* and the member institutions are SUNY-Stony Brook (Project Director: Alan Tucker, Math P137, SUNY-Stony Brook, Stony Brook, NY 11794-3600; e-mail: *LICIL@ams.sunysb.edu*), C.W. Post College, Dowling College, SUNY-Farmingdale, New York Institute of Technology, SUNY-Old Westbury, St. Joseph's College, Suffolk Community College, and York College (CUNY).

all its courses. For instance, in the titration experiment, the students collect data, plot it on the graphing calculator, and locate the point of the inflection of the titration curve, thus showing a nice connection between chemistry and mathematics. The chemistry department reported that "over a short period of time, approximately three semesters, the faculty has noticed a clear decrease in student math anxiety, and an increasing awareness of the importance of mathematics both in and out of the classroom."

Other dramatic increases in student success have been observed in SUNY–Farmingdale's Foundations of Technology course, a nine-credit interdisciplinary course containing mathematics, physics, and technical problem-solving. Students work in small groups using "process education," a teaching methodology developed by Pacific Crest Software. According to the developers, "learning, thinking, problem solving, communicating, assessing, and teamwork are processes to be developed and continually improved by students as they construct knowledge. Process education incorporates cooperative learning, guided discovery activities, journal writing, and various assessment tools" (Krumsieg and Baehr 1996). In the course that used process education, the student passing rate increased from 30 percent to 86 percent.

As part of LICIL's assessment process, evaluation teams visited each of the ten campuses involved in the project. Evaluators have all reported very favorably on their observations, mentioning the enthusiasm of the faculty and the faculty's ability to work with an extremely diverse population. The evaluation teams also noted that making the connections between mathematics and science increased the enthusiasm and level of understanding displayed by the students in the affected areas who were interviewed. Also, anecdotal data indicate that students are doing quite well as they move on to work and graduate school.

It is interesting that many of the LICIL faculty are senior faculty members who remain energized by continuing to change the content of their courses and pedagogy in order to help students, as opposed to engaging in a debilitating daily routine of complaining about the knowledge and skills lacking in "today's students."

In summary, making connections between mathematics and science has led to innovations in content and pedagogy that have enhanced student learning while increasing student interest. Faculty collaboration across the disciplines—and campuses—has been an important component in these successful programs, and one that has enriched the faculty who participate in them.

# Resources to Support the Science Program

**STANDARD D**

> The K–12 science program must give students access to appropriate and sufficient resources, including quality teachers, time, materials and equipment, adequate and safe space, and the community.
>
> - The most important resource is professional teachers.
> - Time is a major resource in a science program.
> - Conducting scientific inquiry requires that students have easy, equitable, and frequent opportunities to use a wide range of equipment, materials, supplies, and other resources for experimentation and direct investigation of phenomena.
> - Collaborative inquiry requires adequate and safe space.
> - Good science programs require access to the world beyond the classroom.
>
> *Source:* National Research Council. (1996) *National Science Education Standards.* Washington, DC: National Academy Press, 218.

Program Standard D is about providing adequate support necessary to ensure a good program. Notice that the single most important resource is a teacher committed to student learning. At the university level, this is the area in which many challenges persist; higher education faculty have commitments unrelated to classroom teaching that are unlike those of teachers in K–12. University faculty must ensure their own promotion and tenure through research; in most instances, research is the primary criterion considered for advancement. Moreover, university professors have entered their profession because of a love of science and because of the opportunities for research through funding. At the same time, it is now apparent to university administrators that undergraduates deserve quality learning experiences—and that these experiences are primarily afforded by faculty who are expert teachers. Thus, many institutions are now seriously considering teaching excellence as a component in faculty advancement. The university professor is finding that an important part of professional development is that required of a scholar-teacher as well as that of a science researcher.

Additional resources required to meet Program Standard D include time for planning—institutions might consider released time for course development as well as for research. The *Standards* call for an inquiry-based program, which is often achieved through laboratory work; but that will necessitate a revamping of many college laboratory programs. Traditional introductory science laboratory programs involve experiments that illustrate and verify concepts presented in lectures; a shift to inquiry-based activities will require a major redesign of the laboratory work, and time must be invested to make the program successful. The *Standards* indicate that real world experiences through fieldwork, internships, and research are essential. Laboratory materials, equipment, and supplies require adequate budgets. Adequate laboratory space is also critical—especially in view of safety issues. University students may be considered as adults who are impacted by the federal Occupational Safety and Health Administration's (OSHA) Emergency Planning and Community Right to Know Act of 1986. Although OSHA's related laboratory standard addresses laboratory employees only, the standard is a good model for providing a safe environment and training for students who are doing laboratory work (OSHA 1990).[2]

These considerations require that a budget be constructed carefully to meet the needs of a quality science program. Particular needs associated with a shift to standards-based teaching will include an examination and refinement of program goals; sufficient faculty to support curricular needs; faculty development in teaching methods; the space, equipment, and supplies necessary for inquiry-based learning; and sufficient money to manage waste and to address the safety issues involved in training faculty, staff, and students. The department or program chair must be a strong advocate of standards-based teaching—and one who knows and understands the research that supports collaborative, inquiry-based courses.

Support for innovation in science education also comes from granting agencies such as the National Science Foundation and corporate foundations. As an example, the contributions of Merck & Co., Inc., are documented in the following vignette.

## From the Field

### Merck Support of Undergraduate Science Education
Susan K. Painter, Manager, Academic Programs
Merck Research Laboratories

Merck & Co., Inc., is a global company composed of over 57,000 people worldwide. Merck believes that an essential component of its corporate responsibility is to provide support to charitable or philanthropic organizations that benefit society—from the local community, where Merck has a major presence, to the international level. Merck makes cash contributions and donates products, technical expertise, and other in-kind services to qualified organizations and programs that address important needs of our society and support Merck's overall business mission to enhance health. Described

[2]Two publications that deal directly with instructional laboratories using chemicals are (1) American Chemical Society (ACS). 1995. *Safety in Academic Chemistry Laboratories.* Washington, DC: ACS, Committee on Chemical Safety, and (2) National Research Council. 1995. *Prudent Practices in the Laboratory.* Washington, DC: National Academy Press.

below is a representative cross section of the various programs we support to further science education at the undergraduate level in the United States.

*Merck/AAAS Undergraduate Science Research Program.* In 1994, Merck, in partnership with the American Association for the Advancement of Science (AAAS), established a competitive grant program—the Merck/AAAS Undergraduate Science Research Program (USRP)—to support undergraduate research. Primarily undergraduate institutions in the northeast and mid-Atlantic states are invited to apply for this grant program. Fifteen awards are made once every three years. The selected schools receive a $20,000 grant annually for three years, awarded jointly to the departments of biology and chemistry to support interdisciplinary undergraduate science research. Merck believes collaboration between these fields is important to advancing scientific knowledge. The objective of the USRP is to encourage students to pursue graduate education in biology and chemistry through undergraduate research experiences at the interface of these sciences. To date, thirty colleges and universities, through the USRP, have provided 403 research experiences for undergraduates. In 2001, the program expanded to become a national competition, with fifteen new grants awarded annually.

*UNCF•Merck Science Initiative.* In 1995, Merck began collaboration with The College Fund/UNCF to establish the UNCF•Merck Science Initiative. The purpose of the initiative is to increase the number of world-class African-American scientists. Currently, less than 2 percent of Ph.D.s in biology and chemistry are held by African-Americans. Merck has committed support to this initiative for ten years with total funding of $20 million.

Through this initiative, thirty-seven fellowships are awarded each year to undergraduate, graduate, and postdoctoral African-American students pursuing careers in scientific research. Recipients are selected from a nationwide competition. This initiative reaches out beyond historically black colleges to all eligible African-American science students in the United States. Fellows receive scholarship or fellowship awards, and each fellow is paired with a Merck scientist mentor who provides valuable research assistance and guidance. Undergraduate fellows also receive internship opportunities at the Merck Research Laboratories for two summers. Institutional support through grants to the science departments of the award recipients is also provided.

*Merck Engineering & Technology Fellowship Program.* A top-tier pharmaceutical company such as Merck needs highly talented engineers to move basic research discoveries into development and manufacture and to overcome many technical challenges with breakthrough technology. Facilities design, construction, and operation are complex and need engineers skilled in advanced processing and manufacturing computer technologies.

To encourage interest of talented students in chemical and other engineering disciplines, Merck launched the Merck Engineering & Technology Fellowship Program in 1997. The fellowship program is available at twenty-one universities in the United States and Puerto Rico. Two Fellows at each school are selected annually from outstanding sophomores or juniors majoring in chemical engineering or other appropriate engineering or science-based disciplines. Each award recipient receives $5,000 to be applied toward academic-year tuition, room and board, or fees. Also, each award recipient receives a paid summer internship at a Merck research or manufacturing facility. Not only do students gain hands-on experience working at Merck, they may also discover a mentor and develop a valuable relationship to enhance their academic or professional careers.

*Pharmacy Programs.* Pharmacy education is another area where Merck provides a variety of types of funding.

Merck gave an educational grant to the American Pharmaceutical Association Academy of Students of Pharmacy (APhA) for the publication of the third edition of *The Pharmacy Student Companion: Your Road Map to Pharmacy Education and Careers.* This book is a quick-reference guide to career evaluation and development resources for the student of pharmacy; it offers practical advice on how to choose and get into the pharmacy school that is right for a particular student. Extensive appendixes contain up-to-date listings of U.S. schools and colleges of pharmacy, scholarships, loans, awards, grants and internships, and state-by-state information on internships and licensure.[3]

An important aspect of undergraduate study in the sciences is presenting experimental results to colleagues and the public. For a number of years, Merck has sponsored an undergraduate research seminar series at the West Virginia University School of Pharmacy, now known as the Pharmacy Student Research Conference–Eastern Region. The daylong seminar is aimed at exposing pharmacy undergraduates to research and graduate education. Students from approximately thirty-five schools located in the eastern region of North America are invited to give paper and poster presentations on their research in the pharmaceutical sciences. This conference has been so successful that Merck has provided support for a similar program at the University of Colorado School of Pharmacy, to be known as the Pharmacy Student Research Conference–Western Region.

Merck provides funding to the following pharmacy organizations to support undergraduate, graduate, and postgraduate pharmacy students:

- The American Association of Colleges of Pharmacy (AACP)— Twelve awards are provided through The Merck Research Scholar Program for students in Pharm.D. programs to conduct research. Awardees are selected

---

[3] The book is distributed to all schools of pharmacy via the American Pharmaceutical Association's Academy of Students of Pharmacy; it may also be purchased directly from APhA (800 237-APHA).

through a national competition.

- The American Foundation for Pharmaceutical Education—Three predoctoral fellowships in pharmaceutical sciences and one post-Pharm.D. fellowship in biomedical research are offered.
- The American Pharmaceutical Association (APhA)—Grants are given to APhA student chapters around the country to conduct

projects to help improve clinical practice.

Developing scientific talent from all sectors of society is in the best interest of Merck, a company whose success depends on innovation. We know that the talent we seek gets its start in the undergraduate science and engineering programs in colleges and universities across the nation.

# Opportunity to Learn

All students in the K–12 science program must have equitable access to opportunities to achieve the *National Science Education Standards*.

*Source:* National Research Council. (1996) *National Science Education Standards*. Washington, DC: National Academy Press, 221.

Occupational data clearly indicate that African-Americans, Hispanics, Native Americans, and Asian/Pacific Islanders are underrepresented in the health professions and as members of the American science research community. Although that may be attributed to lack of encouragement of K–12 students from minority groups to even consider pursuing careers in science, opportunities to become motivated and achieve in science at the undergraduate level are also limited. The lack of services designed to support learning of students from disadvantaged backgrounds and to stimulate interest in scientific fields effectively places a barrier to science as a career possibility to many students from underrepresented groups. This is a great loss to society and to science.

Multiple perspectives enrich science. Our view of the world is broadened when questions asked of science come from scientists with diverse backgrounds. If these questions are not asked, scientists may be slow to consider them. In fact, the rather homogeneous view of scientists as "white males" gives the perception that scientists are a rather elitist group—one that may fail to consider the important questions and the moral and ethical dimensions of its work. How do colleges and universities provide equitable opportunities for *all* students to learn science? The following vignette considers the importance of engaging in research at the undergraduate level to provide motivation and instill an appreciation for science in an underserved population.

# From the Field

### Equity in Science Opportunities at the Undergraduate Level

Elizabeth T. Hays, School of Natural and Health Science, Barry University
National Science Teachers Association (NSTA) Director of Multicultural Education and Equity

How do colleges and universities provide opportunities for all students to meet the *National Science Education Standards* beyond the K–12 science program? Research has shown that the opportunity to *do* science at the undergraduate level is a strong motivator for students who might consider science as a career. Yet the opportunities for undergraduate students—especially the underserved—to have hands-on research experiences are limited. This is particularly true at large research universities, where research opportunities are usually restricted to graduate students; few undergraduates (and fewer minority undergraduates) have such opportunities. Yet smaller universities and colleges have been very active in providing hands-on research opportunities for undergraduates, with much assistance coming from federal programs that have been developed to provide research opportunities for all students.

In the area of biomedical research, the National Institutes of Health (NIH) are committed to increasing the number of minority research scientists. Via the National Institute of General Medical Sciences (NIGMS) and the programs available, colleges and universities with a high minority stu-dent population have been able to develop programs that offer minority students the opportunity to participate in hands-on research. Students perform experiments, work independently on research projects, and publish and present their work at regional and national meetings. Students who participate in these undergraduate programs learn what science research is all about, see their overall academic performance improved by the close mentoring they receive, and find their ability to write and to speak in front of others greatly improved. Their confidence in their ability to do science is so enhanced that many apply for graduate programs at major research universities.

Although these programs are designed to enhance the experiences of underrepresented minority students, all students benefit from the environment that the programs create. Let me now offer some of my experiences with these programs as examples of the hands-on experiences these students receive.

The first summer after I arrived at Barry University in 1984, I was asked to teach the Introduction to Research course for the incoming students in the Minority Access to Research Careers (MARC—now MARC USTAR) program. This NIH GMS program provides tuition, a stipend, and research and travel money for junior and senior undergraduate students who are committed to applying to graduate school in the biomedical sciences. Four Cuban-American students were in my first group. For six weeks, five days each week from nine to five o'clock, we worked together. The students had hands-on experiences in working with and learning the theory of a variety of common analytical research instruments. They performed colorometric as well as spectrophotometric assays. They learned to weigh milligram amounts, to make solutions of various kinds, to keep a research notebook, to determine unknowns. After the six weeks were over, the students were ready to go into research labs at Barry to become involved in projects for the academic year. The summer between the junior and senior year was spent at major research universities such as Purdue and the University of California at Berkeley, Davis, and San Francisco. During the senior year, the students again conducted research on the Barry campus and worked on their undergraduate theses.

In the early years of the program, the students were mainly of Cuban background, but as time passed, the diversity of the students' backgrounds increased, as first Nicaraguans, then students from other Latin American countries, African-Americans, and Haitian-Americans joined the program.

Later, I became a principal investigator in the Minority Biomedical Research Support (MBRS—now MBRS SCORE) program. Here students were paid to work in the research lab and to participate in all the enrichment activities that supplemented the hands-on research component. Although I was supported for only three years by this program, it continues to support the research efforts of many of my colleagues, providing additional opportunities for hands-on research for our undergraduates.

A third program that offers summer hands-on research experience for our students is the Minority International Research Training (MIRT) program. Over the course of six years, six of our students, along with their faculty mentors, have conducted research in Italy, Argentina, and Jamaica.

Research is infectious and it snowballs—a little research leads to more research, which leads to even more research. Science teachers—K–12 and college—should encourage all science students to try it.

# Communities of ——————————————— Support for Teachers

Schools must work as communities that encourage, support, and sustain teachers as they implement an effective science program.

● Schools must explicitly support reform efforts in an atmosphere of openness and trust that encourages collegiality.

● Regular time needs to be provided and teachers encouraged to discuss, reflect, and conduct research around science education reform.

● Teachers must be supported in creating and being members of networks of reform.

● An effective leadership structure that includes teachers must be in place.

*Source:* National Research Council. (1996) *National Science Education Standards.* Washington, DC: National Academy Press, 222.

Communities are encouraged by shared purposes and collaborative work. For example, communities of learners share the need to achieve course goals and will often work together in study groups to meet those goals. Faculty, too, need to work collaboratively to achieve program goals. Faculty who share program objectives form a community in which responsibilities are defined and each member is viewed with respect and fairness.

The Carnegie Foundation for the Advancement of Teaching's CASTL (Carnegie Academy for the Scholarship of Teaching and Learning) program is an example of one effort in building a community of teaching scholars. Called Carnegie Scholars, these faculty members from across the nation receive short-term fellowships and meet with each other as an advanced study group. Examples of the innovations developed by chemistry scholars include the following:

● Investigating student misconceptions in chemical phenomena arising from prior knowledge and experience (James Hovick, University of North Carolina–Charlotte)

● Interventions for at-risk students in large lecture courses (Dennis Jacobs, Notre Dame University)

● Experiential learning activities (Deborah Wiegand, University of Washington)

● Public explanations as a teaching method (Mark Walter, Oakton Community College)

● A professional development program for future faculty that begins with undergraduate curriculum design and ends with the need for new faculty in the project entitled, "Interdisciplinary Studies at the Interface of Education" (Brian Coppola, University of Michigan)

Another component of the CASTL program is the Teaching Academy Campus program, which encourages campuses to build institutional structures that promote local and national dialogues on infrastructure systems that support teaching and learning. An interesting problem arises if you are one of a very few faculty at an institution with an interest in improving learning experiences through standards-based teaching. The following vignette describes how one science department used the services of Project Kaleidoscope to facilitate teacher support and to create a network of reform (PKAL 1999).

## From the Field

### Building Natural Science Faculty Communities
Jeanne L. Narum, Director
Project Kaleidoscope

Project Kaleidoscope (PKAL), an informal national alliance working to strengthen learning among undergraduates in the fields of science and mathematics, began its work over a decade ago with a focus on what works in building natural science communities. It was clear to the members of the alliance that there needed to be visible, tangible institutional support for the ongoing work of developing the policies, practices, and programs that ensure the success of all students. One story from the PKAL experience illustrates the relationship between setting goals for student learning and the institutional capacity to build and sustain strong programs. The story suggests that the fundamental characteristic of community is informed discussion, a process that starts with identifying and asking the critical questions at each stage of the process and exploring them in a context of mutual respect and shared commitments.

The story is that of a comprehensive university in the south, a campus on which the physics department was sharply divided about the need for, or efficacy of, new approaches within the department, even though very few students took more than a single physics course, and the number and quality of majors had been diminishing. At the urging of the small group of faculty in the physics department who were troubled by the current situation and determined to address the problem, the dean of the college invited consultants from the W. M. Keck/PKAL consultancy program to campus for a two-day visit. The consultants, two experienced academic scientists, met with senior administrators, with faculty within and outside the department requesting the consultation, and with students, in addition to reviewing a significant amount of institutional information. Recommendations emerged as the consultants came to understand the problems and potential at the university.

The recommendations began with a reminder that planning and reform take a long time and that eventually there must be a collective commitment to wrestle with educational goals in a way that engages all members of the department and the larger campus community. The recommendations were as follows:

- *Start small.* The consultants urged the faculty to let those who wished to try new approaches with the introductory courses do so, recognizing that there are many ways to teach introductory physics. Teaching introductory courses was a responsibility shared by the departmental faculty; allowing and monitoring various approaches would likely make it clear which of the approaches were the most effective. With that information, more faculty might be empowered to try such approaches. It was important to begin with introductory courses because the most visible evidence of the difficulties within the department was the drop-off in enrollments after the first course: students were voting with their feet. Experimenting with different approaches in the first set of courses was the beginning step to a more flexible curriculum.

- *Have a departmental perspective on the impact and sequence of departmental offerings.* The department was advised to broaden discussions about introductory courses and to think about all courses from the departmental perspective (rather than assuming that individual faculty "owned" individual courses). It was suggested that faculty form teams to examine specific portions of the curriculum and then report back to the whole group. Departmental faculty were encouraged to document the points of attrition and to identify where to make initial changes to maximize the opportunity to recruit students and encourage persistence. In particular, the faculty was urged to look carefully at how their students were making the transition from high school and at the mathematical preparation of incoming students.

This was a faculty with significant externally funded research activity, yet entering students were not given information about the exciting research opportunities that lay ahead for physics majors. Neither the introductory nor the intermediate courses were designed to set students on the path toward a research experience as upper-level students—and turn them into majors. Faculty just did not see students as partners in learning, and students were not being given the opportunity to engage in the kind of collaborative, discovery-based learning that modeled what science is about.

- *Collaborate with and learn from colleagues in other departments.* Oddly enough, this was a campus with a chemistry department actively involved in one of the major NSF-funded chemistry initiatives—a department that had created a carefully sequenced program that provided upper-level students with the opportunity to serve as mentors (role models) in intermediate courses during the academic year and to serve as research partners with faculty during the summer. This program had significantly increased the number and quality of chemistry majors and enhanced the intellectual ambiance and sense of community within the department; clearly, it might serve as an adaptable model for the physics department. Yet, just as there were no conversations within the physics department, there were none cutting across departmental boundaries.

The consultants recommended several curricular avenues to explore with colleagues in other departments, beginning with the coordination or sharing of course offerings by team teaching (with chemistry—thermodynamics or statistical/quantum mechanics; with biology, mathematics, or computer science—electronics, interfacing, or computational methods). Especially since the university had a strong pre-medical and life sciences program, it was recommended that cross-departmental conversations take place about the kind of physics courses required for students in those fields.

At this point, the consultants also addressed the need for visible institu-tional backing, with a set of recommendations to the chief academic officer and the department chairs about supporting activities such as informal conversational lunches (paid for by the dean) to provide venues for faculty to discuss various aspects of the program in a nonthreatening forum. Administrative support for the participation of cross-departmental teams at conferences on new pedagogies and for divisionwide consultants (perhaps with experience in establishing team-taught courses) was also recommended; such activities would bring the experiences and expertise of the larger campus into discussions about transforming individual programs and departments.

These were timely recommendations, in that it was becoming clear that a new science facility would probably be built in the next several years, given the current pattern of planning in the state higher education office. So the consultants also suggested, ever so gently, that attention given early on to gaining broad consensus about the shape of the academic program and future curricular initiatives would serve all departments well as departments went into discussions about a major capital project such as a new building. Speaking from their personal experience, the consultants described how the right approach in regard to facilities planning can assist departments and institutions in defining and redefining their visions for the future.

- *Just start.* One of the final recommendations dealt with the politics of reform, reminding the department that while it is valuable to have total consensus, disagreements can be productive—particularly if they are channeled into opportunities to explore and critique a variety of approaches. Faculty were cautioned about the paralysis that can occur if there is too much pressure for complete consensus before any new approach is tried. They were also discouraged from attempting to force skeptical faculty into trying something they were not prepared to do.

What was the result of the consultation? Most significantly, members of the department attended a weekend-long retreat where they discussed the issues that were dividing the department and established a precise timetable for developing divisionwide student learning goals. Regular planning sessions involved both faculty and senior administrators, all of whom now have a sense of ownership in the process.

## References

Harrill, R. W. (2000) Evolving Curricula in the New Century. *Journal of College Science Teaching* 29(6), 401-407.

Krumsieg, K., and Baehr, M. (1996) *Foundations of Learning.* Corvallis, OR: Pacific Crest Software.

*Mount St. Mary's College Catalog. 1998–2000.* Los Angeles, CA: Mount St. Mary's College.

Occupational Safety and Health Administration (OSHA). (1990) Occupational Exposures to Hazardous Chemicals in Laboratories. Code of Federal Regulations (29 CFR 191.1450).

Project Kaleidoscope (PKAL). (1999) *Steps Toward Reform—Report on Project Kaleidoscope, 1997-1998.* Washington, DC: PKAL.

# Science Education System Standards

| Mario W. Caprio |
|---|

he System Standards of the *National Science Education Standards* (NRC 1996) address standards that go beyond the individual teacher and classroom and apply to the science education system as a whole. The System Standards describe the nature of systems in general as being a hierarchy of subsystems, each with its "boundaries, inputs and outputs, feedback, and relationships" (NRC 1996, 227). To qualify as a subsystem of science education, a unit must influence the delivery of science education either directly or indirectly. This broad view of the science education system underscores the penetration of science into our culture and shows that the stakeholders and participants in science education extend well beyond the traditional community of teachers, scientists, and their disciples.

This section of the *Standards* gives us a general model of the science education system, showing how three enormous subsystems—government, national organi-zations and societies, and the private sector—overlap to influence science teaching (Figure 6.1).

The *National Science Education Standards* are, in a general sense, the set of common values about scientific literacy that have been distilled from a long national debate that constructed our unified ethic for science education. When groups operate within a shared value system, they are capable of more productive action. The System Standards implicitly, but quite clearly, challenge all the system components to communicate more effectively and cooperate more efficiently in the roles they play for science education. The System Standards do this by offering a set of common values and a shared vision under which the separate parts of the system can work collaboratively.

The challenge for improving science education is to find ways in which these enormous systems—and the subsystems they represent—can work together to increase public understanding of science. The seven System Standards, which identify critical issues that need to be addressed by the science education system, are as follows:

**Figure 6.1**

*Interrelationships among Science Education, Government, National Organizations and Societies, and the Private Sector*

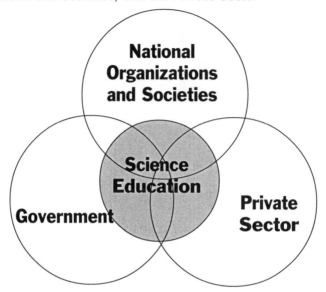

*Adapted from:* National Research Council. (1996) *National Science Education Standards.* Washington, DC: National Academy Press, 228.

**A** Policies that influence the practice of science education must be congruent with the program, teaching, professional development, assessment, and content standards while allowing for adaptation to local circumstances.

**B** Policies that influence science education should be coordinated within and across agencies, institutions, and organizations.

**C** Policies need to be sustained over sufficient time to provide the continuity necessary to bring about the changes required by the *Standards.*

**D** Policies must be supported with resources.

**E** Science education policies must be equitable.

**F** All policy instruments must be reviewed for possible unintended effects on the classroom practice of science education.

**G** Responsible individuals must take the opportunity afforded by the standards-based reform movement to achieve the new vision of science education portrayed in the *Standards.*

*Source:* National Research Council. (1996) *National Science Education Standards.* Washington, DC: National Academy Press, 230–240.

Understanding the system in which our university laboratories and classrooms are embedded is prerequisite to being able to apply any of the System Standards. As noted, Figure 6.1 represents the national science education system and the major subsystems that interact with it as adapted from the *National Science Education Standards*. But in this book, we are narrowing the view toward a more personal perspective at the postsecondary level. Focusing on the university and college level requires some modification of the general diagram. First, we place our science classrooms and laboratories at the center of the diagram; after all, the classrooms and laboratories are where the system produces its product, science literate students. Second, we add one more major subsystem to Figure 6.1—"Campus Resources"—to represent the rest of the campus, exclusive of the science department. Those resources, including libraries, writing and math labs, and psychological and academic counseling services, will influence and be influenced by what is happening in the science departments. The result is Figure 6.2, which presents an overview of the science education system at the postsecondary level.

In a very real sense, the science education system within which each faculty member works has the potential to be an instructional team. To access the power of the System Standards, we must identify the specific components of each of the major subsystems in Figure 6.2; in addition, we need to know how the individual system components are networked with one another and with the science education happening in our classrooms. Following is a brief description of each of the subsystems in Figure 6.2.

## Campus Resources

Of the major subsystems, "campus resources" easily illustrates the concept of an instructional team not usually physically present in the classroom but one whose influence can—when appropriately used—be felt by everyone in the room. In standards-based teaching (see Chapter 1) the instructor serves as a facilitator, guiding student learning and helping students tap into the resources they need to become independent learners. However, the instructor is only *one* resource for scientific information and for student learning. Many college campuses also have reading specialists, writing laboratories, librarians to help with developing information literacy, and math laboratories where students can obtain help with the quantitative aspects of the science courses. With a good campus support system, the teacher of science need not do it all! If students have difficulty with a writing assignment, for example, the science teacher might help them organize their papers and polish their prose; but while scientists may know how to write, most of us only qualify as amateur writing instructors.

Also, these resources come with their own set of electronic accoutrements. Word processors from the writing center, statistical packages and graphing calculators from the math lab, and the Internet search engines and online reference services from the libraries bring powerful new technologies to the student and teacher.

**Figure 6.2**

*Science Education from a Postsecondary Campus Perspective*

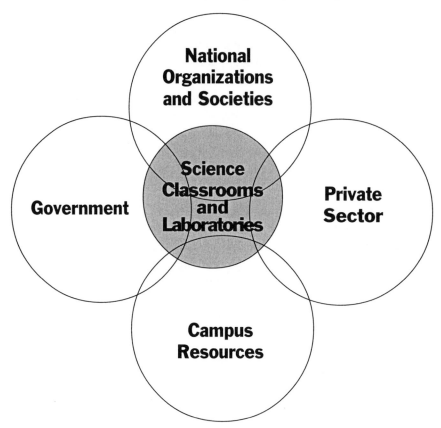

The System Standards urge open dialogue between the system components. Imagine what our students might achieve if science instructors were to work with resource specialists to construct on-campus instructional teams and then maintain team efforts to help our students excel in the math, reading, writing, and even the group dynamics and information literacy necessary to do well in our science courses.

## Private Sector

In the private sector, there are organizations that have a vested interest in science education and that are more than willing to sign on to a more broadly defined instructional team. Private industry can provide field trip sites, specialized instruction, and advice about curriculum (especially in rapidly changing fields). At the undergraduate level, corporations may provide work experiences for students, offer financial support for undergraduate research projects, donate surplus equipment, or sponsor student travel to professional meetings. As we begin to expand our view of the instructional team, we can look to the private sector to forge relationships that will strengthen and broaden what we are able to do in the classroom.

## National Organizations and Societies

National science teaching organizations, such as the Society for College Science Teachers (SCST), the National Science Teachers Association (NSTA), and others with more specific discipline orientation, such as the American Association of Physics Teachers (AAPT) and the National Association of Biology Teachers (NABT), have always focused on issues in undergraduate and K–12 science instruction. In the last ten years, however, there has been a growing awareness of the important role that discipline-based professional societies can play in the improvement of science education. Today, national scientific societies are recognized as the subsystem of science education that brings the perspective of the professional working scientist to the system and for which there is no comparable voice.

In recognition of the importance of professional societies in the reform of undergraduate science teaching, there is an effort underway to facilitate the various societies' contributions to the cause. An organization working in this direction is CELS (Coalition for Education in the Life Sciences), which has as its mission "to improve undergraduate education in the life sciences by bringing the expertise and resources of the life sciences professional societies to bear upon critical issues relating to life science undergraduate education in the United States" (CELS 1998).

## Government

Government agencies informed by the scientific community, the nation's colleges and universities, and the general electorate decide how tax money is to be spent in support of science education. Those decisions determine, to a large extent, the direction that science education will take. It is very important that funding support the priorities set by collaboration among all parts of the science education system, including the local, state, and federal governments. An example of the federal government promoting the System Standards came through the "Shaping the Future" conference sponsored by the National Science Foundation (Rutledge 1996); this conference is a model for collaboration between the major subsystems of science education.

To access the power of the System Standards, we must be able to identify the specific components of each of the major subsystems described above, and we need to know how the individual system components are connected to one another and with the classroom.

There is likely to be fairly uniform agreement among colleges and universities about the four subsystems of science education as they appear in Figure 6.2. We expect, however, that a list of components under each major subsystem will vary in kind as well as in significance among institutions. Table 6.1 lists some specific components of the major subsystems in postsecondary science education. The more specificity we can bring to each list of components, the richer will be our understanding of the local system and the more useful the list will be for us. However, it is also quite clear that in order for a very specific list to accurately reflect the science education system at a particular institution, it needs to be constructed locally.

Table 6.1 is presented as an example of how to identify the relevant science education subsystem in one's institution. Making such a list is an important first step when a college, academic division or department, or a professor is ready to align with the System Standards of the *National Science Education Standards*.

**Table 6.1**

*Examples of Components of Science Education's Major Subsystems*

| Campus | Government | Nat'l. Orgs. and Societies | Private Sector |
|---|---|---|---|
| Computer laboratories | Federal, state, and local legislative bodies | National Science Teachers Association | Area hospitals |
| Math lab | U.S. Department of Defense | Society for College Science Teachers | Industrial laboratories |
| Writing center | National Science Foundation | American Chemical Society | Private foundations |
| Counseling departments | U.S. Department of Energy | American Association of Physics Teachers | Museums |
| Libraries | Parks and wildlife refuges | Various discipline-based organizations | Local radio and television stations |
| Professional development centers | Resource management agencies | | Chambers of commerce |
| Board of trustees | State education departments | | |
| Academic departments | Local school districts | | |

## Constructing a Local Model System

In addition to listing the components of the major subsystems and showing how they connect to one another and with science education, our representation of the system would be more useful if it indicated the relative weight that each of the system components brings to bear on science teaching. For example, the funding carrot held out by a granting agency certainly helps to move that agency's agenda for science education and can sometimes encourage faculty members to adjust their direction and align their professional goals with that of the funding source. But funding cycles and the success rate of acquiring funds puts the influence of the funding agencies at a much greater distance from the science classroom than that of the academic department, which teachers will interact with on an almost daily basis and which is where issues such as peer pressure and decisions about promotion and tenure reside. For purposes of describing the system, we might say that academic departments are *close* in their influence on the classroom, while funding agencies are more *distant*. Figure 6.3 attempts to show how the individual system components are related to one another and to the classroom as well as how information and resources flow between the various system components. As in Figure 6.2, the diagram has the science classrooms and laboratories at its center. Then, with some subsystems being closer or more distant in terms of their impact on the classroom, the chart begins to look as if it will suggest concentric spheres of influence into which we can position the various system components. Finally, arrows indicate the direction information and resources flow between the components. A typical diagram of the education system at a specific university might be represented by Figure 6.3; however, the details of the diagram will differ, depending on the characteristics of each university.

**Figure**

*Typical Diagram of a University Education System*

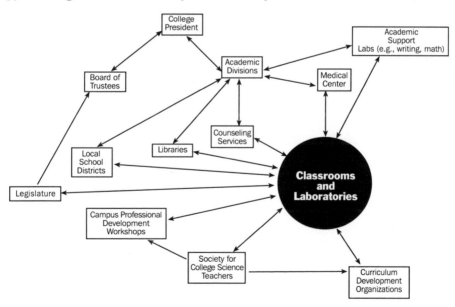

*How would you change Figure 6.3 to better represent the science education system from the perspective of your campus?*

Variation in local higher education systems is expected and is consistent with the *Standards* for two reasons. First, the *Standards* are not prescriptive; they explicitly allow for local differences. Second, the *Standards* are a set of values that, in the case of the System Standards, do not pretend to list the system components with any specificity. Instead, they talk about the relationship between the components and offer guidelines to help optimize their interactions with one another. We have gone through the exercise in this chapter to demonstrate the procedure that a college, an academic division or department, or a professor needs to follow as a first step in aligning practices with the System Standards of the *National Science Education Standards*.

## Sharing a Vision

Policies that influence the practice of science education must be congruent with the program, teaching, professional development, assessment, and content standards, while allowing for adaptation to local circumstances.

*Source:* National Research Council. (1996) *National Science Education Standards.* Washington, DC: National Academy Press, 230.

"If the practice of science education is to undergo radical improvement, policies must support the vision contained in the *Standards*" (NRC 1996, 230). System Standard A says that the various categories of *Standards* work together as a unit. Developing policies or procedures that conflict with any of the *Standards* will adversely affect the effectiveness of science education.

The *Standards* proclaim teaching to be a highly professional field with an enormous body of knowledge to support it and an ongoing research program to continue its advancement. To be a teacher and display the degree of excellence that the *Standards* set in the sections on Teaching, Assessment, and Professional Development is a full-time career commitment. The *Standards* strongly support a scholarship of teaching and to be truly aligned with them means acceptance of that premise on some level. But, even if they were interested in it, many people see involvement in the scholarship of teaching as too risky. Some fear, for example, that it will take them into new and unfamiliar territory that will deflect time from scholarship in their chosen fields, result in reduced scholarly output (publication), diminished status in the eyes of their colleagues (peer review), and threats to their future support (grants). New faculty members fear that too much emphasis on teaching may jeopardize their chances of securing tenure—a not unjustified fear given past experience (Middleton 1997).

Until recently, excellence in teaching at many colleges and universities was considered nice but not necessary; the real importance was placed on the faculty member's research and publication in the discipline. Promotion and tenure decisions rarely considered a faculty member's scholarship in teaching. Indeed, science professors who published educational research were often told that publication in educational journals, even though they may be refereed, didn't really count. And horror stories abound of teachers who have been denied tenure or who have even been dismissed because they were committed to *Standards*-based innovative approaches in a "telling is teaching" environment. System Standard A says that "policies that influence the practice of science education"—and policies of promotion and tenure would seem to qualify as influencing science education—"must be congruent with the program, teaching, professional development, assessment, and content standards..." (NRC 1996, 230). But change is underway! Even in large institutions noted for research and graduate programs, good teaching at the undergraduate level is becoming an important and integral part of the faculty review process.

The University of California (UC) is addressing *Standards*-based teaching in its undergraduate programs. At a conference in spring 1997, key science faculty and administrators from its eight campuses met to identify "the role of the University in implementing the Standards, and the impact that the Standards will have on University-level science education, particularly at the lower division level" (UC 1997). There are many national groups that deliberate issues in undergraduate science education, yet only a handful of faculty and administrators routinely cross boundaries between the groups. The American Chemical Society's Division of Chemical Education, one of the oldest organized divisions in a large professional society devoted to teaching and learning, held a symposium in spring 2000 dedicated to the need for greater articulation and participation among some of these groups. Representatives of several groups participated, and the following vignette summarizes some of the remarks made.

## From the Field

### Strength in Numbers: Uniting the Fronts of Higher Education (Summary of Symposium)

Brian P. Coppola, Department of Chemistry and Symposium Organizer
University of Michigan, Ann Arbor

This symposium, held during the 219th National Meeting of the American Chemical Society, March 26, 2000, in San Francisco, California, featured representatives of the Carnegie Foundation for the Advancement of Teaching, the American Chemical Society (ACS), the American Association of Higher Education (AAHE), the National Association for Research on Science Teaching (NARST), and the Gordon Research Conferences (GRC).

Lee Shulman, president of the Carnegie Foundation for the Advancement of Teaching (*www.carnegie foundation.org*), reflected on the impact that the 1990 publication of Boyer's *Scholarship Reconsidered* had had on institutions, regardless of their size and specialization. He expressed a degree of frustration, however, that the scholarship of teaching concept has not been more fully integrated into a broader understanding of scholarship. When one reviews a research portfolio, he noted, one has access to a real body of evidence on the quality of a mind. This impressive process involves at least three rounds of review: when the work is proposed, when the publication is reviewed, and when the dossier is examined. When you take scholarship seriously, you make your work public, you are eager to submit it for review, you put it in a form that can be exchanged and simply given away, and you hope and expect that others will build on it. The Carnegie Foundation for the Advancement of Teaching is focusing its effort on understanding what it will take to allow teaching and learning the same degree of dignity and respect as a form of scholarship. We need to move our assessment of teaching, exhorted Shulman at the symposium, past a measure of whether or not the instructor is safe to be left alone in a classroom. Teaching must be as intentional as research, and it must be able to be documented in such a way that it can become public, shared, and built upon.

Shulman and his colleagues have a sense of what the scholarship of teaching and learning will entail in a broad sense, but they also know that insights and particulars will come from within the disciplines. He described the three parts of the foundation's CASTL program (Carnegie Academy for the Scholarship of Teaching and Learning). The Carnegie Scholars are a group of faculty from across the nation who are provided with short-term fellowships and the opportunity to meet with each other as an advanced study group. The Campus Teaching Academy, another Carnegie program, is encouraging campuses to build institutional structures that will promote local and national dialogues on infrastructure systems that support teaching and learning. Finally, CASTL seeks to work with the disciplinary societies and associations because there will not be universal processes, but rather highly idiosyncratic solutions that reflect the disciplines themselves.

Jerry Bell, coordinator of graduate-level program activities of ACS, provided an overview of the five areas of interest to the society. Those areas are (1) the graduate school experience (e.g., workshops, publications, and fellowships and awards); (2) data collection (e.g., recent large-scale surveys that have studied the character of the work force, employment and salary patterns, and TA training activities); (3) policy issues (e.g., white papers, task forces, conferences, and congressional testimony); (4) consciousness-raising activities (e.g., presidential events at ACS national meetings and a variety of invitational events and publications); and (5) career information (e.g., workshops and related events and the partnership of the ACS with the Association of American Colleges and Universities, the Council of Graduate Schools, and the National Science Foundation in supporting the national Preparing Future Faculty program. (For more information on the Preparing Future Faculty program see the From the Field vignette for Professional Development Standard D or *www.acs.org* and *www.preparingfaculty.org.*)

David Malik, chair of chemistry at IUPUI (Indiana University Purdue University Indianapolis) and Ivan Legg, provost at the University of Memphis, both reflected on their experiences with AAHE. A number of changes in higher education over the past decade can trace their roots to AAHE activities, namely, the peer review of teaching, the use of pedagogical colloquia in hiring, and revised assessment and promotion guidelines according to the scholarship of teaching and learning. Malik and Legg both pointed out that significant discussions take place at ACS–AAHE meetings about supporting the whole faculty member and the often-difficult transition from graduate school. A highlight of AAHE meetings is two to three days of workshops that provide many practical lessons. Both of these administrators have supported new promotion and tenure guidelines that take seriously a broadened understanding of scholarship. (For more information, see *www.aahe.org.*)

Mary Nakhleh (Purdue University) is an active member of NARST, the National Association for Research on Science Teaching. A venue for multidisciplinary communication, the postsecondary representation at NARST has grown steadily since 1990. In 1998, the national meeting began to support a full strand of sessions devoted to postsecondary teaching and learning. NARST also hosts a large contingent of international researchers at the national meeting, thereby bringing an important global perspective to higher education. (For more information, see *www.narst.org.*)

Barbara Sawrey (University of California, San Diego) reviewed the history of the Gordon Research Conferences (GRC). In 1993, the first and only GRC devoted to K–16 science education was offered in Ventura, California. Starting in 1995, the topic was shifted to Innovations in College Chemistry Teaching. The GRCs are small meetings in small venues, where everything is off the record in order to promote candor and free exchange of information. Speakers are arranged by invitation only, and attendees need to apply to participate. GRCs are designed to be crucibles where open, multidisciplinary, state-of-the-art discourse is used to move ideas forward. (For more information, see *www.grc.uri.edu.*)

In the afternoon session, Pat Hutchings, senior scholar at the Carnegie Foundation, provided more detail on the CASTL program. She described the concept of "strategic incrementalism" proposed by Larry Cuban (1999) in his book *How Scholars Trumped Teachers* as a model for promoting institutional change. ("Strategic incrementalism" is a succinct way of describing change that occurs in little bits, all in the same direction.) One of the desired outcomes of bringing the efforts of higher education together, Hutchings says, is to see teaching as a way of creating and building knowledge.

Activities of each of these groups support the principles articulated in the *National Science Education Standards*; through coordinated efforts, program innovations that lead to enhancement of undergraduate teaching and more effective learning are becoming widespread.

## Coordination of Science Education Policies

Policies that influence science education should be coordinated within and across agencies, institutions, and organizations.

*Source:* National Research Council. (1996) *National Science Education Standards.* Washington, DC: National Academy Press, 231.

System Standard B, with its emphasis on communication and a shared vision for science education among components of the system, offers a mechanism to effect change. Once again we see the *Standards* speaking to the need for a common value system within which the proponents for scientific literacy can operate efficiently and effectively.

When different institutions share a vision of course content, agreeing on what knowledge and information students are expected to have upon completing specific courses, it makes course transferability decisions more accurate. More and more colleges and universities are entering into articulation agreements to make it easier for students to move between institutions. In general, those agreements are arrived at by looking at existing courses and determining where equivalents exist. If the institutions involved were closely aligned with Standard B, they would have a shared vision of the content of introductory biology, for example, and—working independently—would have constructed parallel courses in the first place. In a sense, this would result in articulation agreements being constructed from the bottom up.

Coordination based on common values does not mean rigid uniformity. Knowledge of basic principles of ecology may be expected of any student exiting an introductory biology course, but a school on the Maine coast might achieve that content goal by studying the marine environment, while the same course taken in Oklahoma might prefer to achieve the goal by studying a different biome. The *Standards* never suggest that introductory biology will be identical across the country.

In fact, neither introductory biology nor any other college-level course is mentioned in the *Standards*. Instead, the *National Science Education Standards* address the K–12 spectrum, and they talk about content by grade level. The implication for colleges and universities is that content standards exist for them too (refer to Chapter 4 of this book for a more thorough discussion of content issues), and the value of different institutions agreeing on course content goes well beyond facilitating course transfers. When all institutions agree on the knowledge of science that a nonmajor ought to have after completing a certain program of study, a cultural norm for scientific literacy in the nation is established.

System Standard B also has implications for the establishment of in-house budgeting priorities and funding of institutional initiatives by outside agencies. Again, when cooperating subsystems have common values and a shared vision of where science education needs to go, it helps ensure that resources (this usually means money) will be focused on the goal. It is reasonable to expect that as more funding agencies adopt the national vision of science education, project proposals aligned with the *National Science Education Standards* (or the collegiate implications of those standards) are more likely to be funded. On the college and university level, this might mean that campus administrators who control funding to academic departments are more likely to support work that is part of a coherent national effort.

The coordination of policies requires continuous communication, first to establish those policies and then to adjust them in light of experience. Inasmuch as the *Standards* are the common national vision, the policies themselves are expected to align with the Content, Teaching, Program, Assessment, and Professional Development Standards. If that is the case, policies developed by one organization or institution are not likely to work at cross-purposes to those created by another group. The implications of System Standard B are far-reaching.

Ongoing communication between postsecondary institutions and outside organizations that are part of the science education system can develop and maintain a *Standards*-based, shared vision of science education that will coordinate goals and facilitate support of programs to advance scientific literacy. These outside organizations are not limited to funding agencies. They include textbook and other curriculum materials publishers, legislative bodies, and industrial laboratories that accept student interns. And, from the *Standards*-oriented classroom teacher's perspective, communicating with other members of the academic department to ensure that a shared vision exists between them is imperative if teacher-initiated innovation is to receive the recognition and support it needs to be successful.

In the following vignette Doug Schamel and Leslie Gordon describe an innovative approach to teaching science to elementary education majors (Schamel and Gordon 1997). The program is interesting in and of itself, but it also illustrates the interplay of several system components in the process of achieving the goal, and it shows what can happen when system components are *not* operating with the same values. If all the components involved here had subscribed to System Standard B, had been in close communication, and had been aligned with the *National Science Education Standards*, the story would have had a very different ending.

# From the Field

## Cost-effective Biology for Elementary Education Majors

Douglas Schamel, Department of Biology
Leslie Gordon, Department of Education
University of Alaska, Fairbanks

## WANTED:
### Science Courses Designed for Would-be Elementary Teachers, the Largest Major on Campus.

This sign should have been in the window of our administration building years ago. The University of Alaska, Fairbanks (UAF) had specialty courses for education majors in mathematics, music, and art, but not science. Elementary education majors in introductory biology laboratories often asked how they could recast science concepts learned at the college level into a more appropriate form for elementary school students. Accomplishing such a transformation is not easy. It requires fundamental understanding of the science concepts combined with knowledge of age-appropriate terminology and processing ability. Although one-on-one conversations can address some specific concerns, these brief sessions cannot substitute for extended training and practice. It simply takes time for students to get the knack of transforming these concepts. As a result, part of Schamel's sabbatical leave mission included investigating science courses geared specifically for elementary education majors, with the intent of developing such a course at UAF.

### In the beginning

Rushing in to see the department head with the blueprints in hand for the new course, we immediately were met with a blast of cold water: It was not *really* a biology course. Oh, and by the way, there was no money to create a new course, anyway. Not taking *no* lightly, we moved on to the dean's office: great idea; sounds exciting; who is going to pay for this? We have no money! Next was the School of Education: Wow, what a great idea! Sure, our students would be happy to enroll. Money? There is no money. But we still think you should teach it.

The wall seemed too high to climb over and too wide to walk around, but Schamel and a friend teaching part-time in education and part-time in biology devised a plan. Since the two of us were slated to co-teach the nonmajors' introductory biology course, we could dedicate one lab section for elementary education majors. That section would have different expectations and materials. The elementary education students were enrolled in this course anyway, and we were teaching it anyway. So who would care?

At the last second, the part-time faculty member was hired full-time to teach education courses and had to back out of co-teaching the biology course. Fortunately, she assisted with the development of the laboratory. Since these students would eventually enroll in her Methods of Science course, she had a vested interest in the success of this special laboratory section.

The paradigm for the lab was the Alaska Science Consortium Learning Cycle Model. We wanted students to feel comfortable investigating questions. And we wanted them to be so well versed in the model that they could build lessons in their sleep. We also wanted them to be proficient in science process skills—the tools of the trade. So, as often as possible, we worked them through observations, inference, and graphing. And we knew they needed to have experience with basic equipment, so we used hand lenses and microscopes, test tubes and hot plates. The emphasis was on easy-to-obtain materials, some of which we had them make.

### The course evolves

A year later, the course was again scheduled. This time Schamel teamed with Leslie Gordon, a third-grade teacher who had been active in teacher training and who had received recognition for excellence in science teaching. It was a perfect match. Schamel brought scientific expertise and some pedagogical skills, and Gordon was steeped in exemplary pedagogy, while being adept in scientific inquiry.

The addition of Gordon to the instructional team added new, and powerful, dimensions. A frequent comment during labs was, "This is a great experiment for us, but I don't think a third grader could handle this concept." Gordon's reply often was, "I did this very exercise with my class last month. They not only loved it, but have gone beyond it to work on X, Y, and Z. One student even did an 'I-search' paper on an extension and is now starting a science project." Our students were immediately convinced, because it had happened—and happened successfully—at the elementary level. They quickly learned not to underestimate what a motivated elementary school student is capable of learning.

Gordon's membership on numerous advisory panels meant that she attended frequent meetings. Students were briefed on "what's new on the science education front" upon her return. With the development of new curriculum materials, we added new modules into our labs.

### Coordination with the education department

Since the faculty member in the education department had breathed life into this course, it was not difficult to coordinate with that department. We asked for, and received, an updated "shopping list" each year: topics or concepts she wanted us to cover. This would ease the transition between the two departments, and save her time, so she could move on to more topics. A nice touch was her visit to our class to welcome these students into the world of elementary education and to assure them that time spent on our assignments would pay off in her course. She specifically mentioned the Learning Cycle Model, and watched for recognition. In return, our students beamed, because they felt comfortable with the model.

### Problems in paradise—The beginning of the end

Just when we thought the sea was smooth and the sailing would be easy, storm clouds began to appear with frightening speed.

- The cooperating faculty member in education resigned to accept a position elsewhere.
- Schamel was reassigned to a different course, but retained control of the Education Lab.
- Money for teaching assistants became tight, as research assistantships became scarce.

No one was rehired to teach the methods course on a permanent basis. Thus, our support in the education department dwindled. The faculty member in science who took over the nonmajors course combined lecture and laboratory material on hour exams, and students in the Education Lab were disadvantaged, because they were not learning the specifics delivered in the "regular" labs. The biology department wanted to use money elsewhere for graduate student support, so our course failed.

Students are now left with no science course to help translate from the college to elementary level. The focus of the teaching methods course is the *teaching* of science, and the instructor assumes students walk in with a science background. There simply is no time there to learn all the concepts we integrated into our biology lab. And there is no easy way for the students to begin to reflect on teaching science while they are enrolled in a content course.

Our fatal mistake was to focus too much on the needs of our students and not enough on the effective coordination needed to sustain the course. We should have spent more time building a fan club in the education department. We should have convinced the dean of natural sciences that a service course, such as this one, was vital to the mission of the institution as a whole. We should have mandated money for the continuation of the course, shared equally by both departments. We should have asked the local school district to write letters to the university, praising the course and requesting more like it. Yes, we focused too much on teaching, and too little on politics. But, then again, that's who we are—teachers.

*The sign is back in the window.*

**WANTED:**
Science Courses Designed for Would-be Elementary Teachers, the Largest Major on Campus.

## Sustained Policies

Policies need to be sustained over sufficient time to provide the continuity necessary to bring about the changes required by the *Standards*.

*Source:* National Research Council. (1966) *National Science Education Standards.* Washington, DC: National Academy Press, 231.

System Standard C recognizes that it takes time to make significant changes in social systems and that the time frame for educational reform may not necessarily coincide with the length of a political office or the duration of an academic administrative appointment. Changes in administrations are not sufficient reason to terminate ongoing innovative programs before they have matured.

On the level of classroom teaching, Standard C reminds us that it takes time to learn. This is true of instructors who are learning to master new classroom approaches and of their students who are grappling with new subject matter. Skipping from one teaching approach to another, deciding that a change is imperative after only one semester, reflects the impatience of inexperience. We have heard this sort of remark all too often: "You know, I tried collaborative learning last semester, and it didn't work for me at all, so I decided to go back to lecturing—it's what I do best."

But it took time for that teacher to learn how to lecture. It certainly takes more than a semester for anyone to begin to feel comfortable behind the podium of a lecture hall, and it is important that a realistic amount of time be devoted to mastering standards-based teaching, too. But having patience for mastery is not just an admonition to the classroom teacher. Teachers who willingly move in the direction of reform are taking a bold step. They are departing from what has typically earned them the approval of their supervisors and peers, and they don't know for sure how quickly or how well they will be able to make the shift. Significant professional development may be required for the faculty member to master the new approaches, and it will take time for the supervisor to master the formative techniques that can be helpful to the faculty member in transition. When academic administrators and teaching faculty all understand and share the values and vision of the *Standards*, however, the transitions are less stressful for everyone (see System Standard B).

# From the Field

## Nurturing Meaningful Relationships in Science Education

Stacy Treco, Director of Marketing
Benjamin/Cummings Science, an Imprint of Addison-Wesley Longman

Education is about partnerships: educator to student, author to reader, science publisher to the science community. As an educational publisher, Benjamin/Cummings sees the partnerships with authors, contributors, reviewers, professors, and students as an essential part of our contribution to the educational process. One partnership in particular, the *Strategies for Success* series, illustrates our commitment to science education.

Since 1988 Benjamin/Cummings has hosted an ongoing series of workshops called Strategies for Success and produced a newsletter of the same name. The workshops are held approximately six times a year at college campuses across the country. The newsletter is published three times a year, with back issues available on our Web site at *www.awl.com/bc*. Both are intended as a service for undergraduate science instructors to stimulate ideas, discover solutions to common obstacles, and provide updates on recent developments and findings.

The *Strategies for Success* newsletter is mailed to 20,000 instructors three times a year, with thirty issues published to date. Each issue explores a particular theme and contains information about upcoming conferences and special events. For example, a re-cent issue (fall 1997) focused on building student participation and featured an article by Ruth E. Beattie, University of Kentucky, entitled "The Large Enrollment Classroom: Creating a Participatory Learning Environment." In that article Professor Beattie, who teaches introductory biology courses of approximately three hundred students, discussed her participatory strategies and perceived barriers to them. She outlined two of the most successful strategies—Think-Pair-Share and One-Minute Papers—which give students the opportunity to participate actively in the course but don't consume too much class time. (In a Think-Pair-Share activity, the instructor poses a question to the students: What do you think would happen if…? What is your opinion about…? Students work alone on the question for one minute and then share their answers with the student seated next to them. A few volunteers then share their answers with the whole class. When using One-Minute Papers, the instructor—near the end of a class—asks students to write for one minute on a given topic, which may be an analysis of a critical thinking problem or comments on areas of the day's lecture that were unclear or confusing to the student. The instructor reads the papers before the next class period, at which time he or she presents a summary of the contents of the One-Minute Papers and reviews any problem areas that have been identified.

The lead article in another recent issue of the *Strategies for Success* newsletter posed an intriguing question: "Adaptation is a familiar topic in the life sciences, but can we apply the concept to the curriculum itself?" Adapting to changing students' needs is what *Strategies for Success* has always been about. For example, one of the issues facing professors today is the struggle to integrate technology into the classroom. Through the workshops and newsletter, Benjamin/Cummings has helped instructors follow the path from videodisks and barcode readers to PC/Mac disks, to CD-ROMs and Web products

The workshops—ninety to date—are a valuable forum. They were created to assist the many instructors unable to attend national meetings and conferences due to cost and travel considerations. *Strategies for Success* offers them a meeting place, close to home, where they can enhance their professional skills and share their teaching experiences and concerns with colleagues. The workshops are offered free of charge and cover a broad range of topics in biology, anatomy and physiology, microbiology, and chemistry that closely match the interests of science teachers. At a recent workshop at Jefferson State Community College, in Birmingham, Alabama, the hands-on and lecture presenta-

tions were Cyberspace: Another Dimension to Teaching the Sciences; The Large Enrollment Classroom: Creating an Inclusive Learning Community; Creative Teaching Strategies in the Sciences; The Grant Game: A Few Simple Rules/Several Bad Ideas; and Hands-On Multimedia.

The workshops attract a mix of experienced educators committed to sharpening their teaching skills, as well as those new to the profession. The sharing of ideas is a cornerstone of the workshops' success. A workshop held at Anne Arundel Community College in Arnold, Maryland, also featured a first-time offering of a chemistry track that continues to develop and evolve. The sessions at that workshop were Connecting the First Day, Inquiry-Based Instruction, and Problem-Solving in Chemistry.

The most important component of the *Strategies for Success* series is the participants themselves. They come to the workshops, read the newsletter, and contribute their own articles because they care about their students and are eager to explore the ways in which they can become better, more informed teachers. Many are struggling with issues that may not have even existed when they began teaching: "How do I incorporate media into my classroom?" "What is the best way to present a difficult topic like cellular respiration?" Others are intent on overcoming more fundamental problems: "How do I capture and hold my students' interest?" "How can I turn my great idea into a 'fundable proposal'?"

The goal of Benjamin/Cummings' *Strategies for Success* series is twofold: to develop a lasting relationship with science instructors that will assist them in becoming better teachers and to gather feedback that will help us continue to create valuable teaching tools that meet their needs and the needs of their students. We are pleased to provide a forum where educators, authors, and publisher can meet to develop the tools that will take science education into the new millennium.

## Resources for Change

**Policies must be supported with resources.**

*Source:* National Research Council. (1996) *National Science Education Standards.* Washington, DC: National Academy Press, 232.

The reform of science education requires an effort by all parts of the system that goes beyond what is required for traditional teaching. It takes work to effect change, and while that work is being done, there is still the day-to-day business of teaching students, attending meetings, providing tutorial assistance, and carrying out the other routine tasks of teachers and administrators. If resources are not made available to energize the reform movement in this already busy environment, reform will not happen.

The resources needed for change vary, but all of them can ultimately be translated into dollars. Some of the ways in which these dollars may be spent will support the following:

- *Professional development programs.* Changing the way we teach students must be supported by professional development programs. Who will pay for those programs and who will pay the teachers to attend them?

- *Inquiry sites. Standards*-based science teaching emphasizes inquiry; an inquiry approach means every science course needs an appropriate inquiry site. This usually means a laboratory, but there are some excellent computer simulations that can also achieve this end. In either case, wet lab or computer lab, the *Standards* do not support no-lab science courses.

- *Information systems.* As students become independent learners, they become more information literate and seek sources beyond their textbooks. Are our libraries and other campus information systems adequate to support this ramification of *Standards*-based teaching?

As a group, higher education faculty are dedicated people. They will often give their time and go above and beyond the job requirements for a professional ideal, and they are sometimes taken advantage of because of that. Recall the response of the education department to Schamel and Gordon (above): "Wow, what a great idea! Sure, our students would be happy to enroll. Money? There is no money. But we still think you should teach it." And they did.

There are many examples of this sort, where faculty did what was right for their students without system support. The problem is that all these efforts take time, and science teachers and science educators are people who, like most other people, have mortgages to pay, orthodontists to support, and automobiles to repair. Among other things, uncompensated use of their time saps their energy for official responsibilities, robs their communities of their participation in civic matters, and decreases the time they have with their families. It is no wonder that uncompensated professional activities are generally unsustained. The reform effort in science education is just too important to load with such uncertainty.

The following vignette illustrates how a professional society can assist in providing resources for change at the undergraduate and graduate level.

## From the Field

### Resources for Change: Educational Activities of the American Chemical Society

Stanley H. Pine, Professor of Chemistry
California State University, Los Angeles
Past Chair, ACS Committee on Education

The American Chemical Society (ACS), through its Office of Education and International Activities, continues its long-standing leadership in chemistry and science education. Nationally chartered to advance chemistry in all its branches, including education, the ACS is the world's largest scientific society, with over 159,000 members. Its 1997 statement "Science Education Policies for Sustainable Reform" is a seventeen-page booklet outlining a broad scope of educational goals. The ACS was involved in the development of the *National Science Education Standards* from the beginning of this National Research Council–guided project. The ACS Task Force on the *National Science Education Standards* had a continuing dialogue on the project, and Sylvia Ware, director of education at the ACS, was a member of the National Science Education Standards committee. Shortly after the release of the *Standards*, the ACS developed *Chemistry in the National Science Education Standards*, a reader and resource manual for high school teachers. This book, which has been widely distributed throughout the United States, provides perspectives and examples to help teachers integrate the goals of the *Standards* into their curriculums.

The ACS is continually promoting new ways of teaching and learning through its educational materials. Learning goals are critical to its curriculum development. And the ACS believes that active hands-on, as well as minds-on, experiences best serve students at *all* levels in learning science. A variety of ACS educational resources are available to science faculty and teachers, and the ACS Education Web page, part of *www.acs.org*, provides continually updated information on these materials. As new resources are produced, appropriate references to the *Standards* are included.

Technician education, which commonly begins in the last two years of high school and continues through community college, is a high visibility area of ACS education. Through its NSF-funded *SciTeKS* project, ACS is developing an extensive curriculum to address the needs of these students. Technician education curriculums are further guided by the Foundations for Excellence in the Chemical Process Industry project, an extensive ACS project funded by the U.S. Department of Education that enumerates and illustrates the specific requirements of technicians in industry.

The extensive efforts of the ACS to enhance undergraduate education affect those students who will become teachers, and thus, they affect K–12 teaching. Teachers typically duplicate their college learning experience in their K–12 classrooms. One of the most successful chemistry courses for nonscience major undergraduates is based on the ACS textbook, *Chemistry in Context*. Now in its third edition, this curriculum promotes chemistry learning through science-related issues that influence all of our lives. It is a wonderful example of active, contextual-based learning.

Many activities of the ACS undergraduate Student Affiliates program relate to outreach and breadth in the learning experience and thus have a direct impact on those undergraduates who have an interest in teaching. Various experiential opportunities provide learning beyond the classroom. A special grant program, Community Interaction—Student Affiliates Grants for Affiliate Intervention, is designed for outreach at the K–12 level within the community.

The ACS is also playing an active role in the National Science Foundation portion of the well-established Preparing Future Faculty (PFF) Program of the Association of American Colleges and Universities and the Council of Graduate Schools. This effort affects the graduate education of students who intend to have academic careers, and thus, it affects the future faculty of undergraduate students. The program recognizes that although college faculty seldom have any formal training in teaching, future teachers will model their own teaching patterns and goals on what they see in the college classroom.

In addition to the many education resources provided by the ACS, the society also plays another very important role: that of an advocate for science education. The ACS is highly regarded by the members and staff of the U.S. Congress as well as by the legislators of many states. The ACS regularly monitors and comments on the status of legislation related to education and promotes modes of funding appropriate for that mission. The ACS is a resource often called upon by decision makers.

## Equitable Policies

Science education policies must be equitable.

*Source:* National Research Council. (1996) *National Science Education Standards.* Washington, DC: National Academy Press, 232.

A guiding principle of the *National Science Education Standards* is that science is for all students, and equity is stressed in each of the *Standards*:

- Teaching Standard B: "Teachers…recognize and respond to student diversity and encourage all students to participate fully in science learning" (32).

- Assessment Standard D: "Assessment practices must be fair. Assessment tasks must be reviewed for the use of stereotypes…. Large-scale assessments must…identify potential bias among subgroups" (85).

- Content Standards: "…all students should develop understanding and abilities…" (115).

- Program Standard E: "All students must have equitable access to opportunities…." (221).

The equal opportunity stressed in the standards above will fail if the system policies are not equitable—and our science and our society will be the poorer for it.

The statement describing System Standard E applies to all levels of education:

> Equity principles repeated in the introduction and in the program, teaching, professional development, assessment, and content standards follow from the well-documented barriers to learning science for students who are economically deprived, female, have disabilities, or are from populations under represented in the sciences. These equity principles must be incorporated into science education policies if the vision of the standards is to be achieved. Policies must reflect the principle that all students are challenged and have the opportunity to achieve the high expectations of the content standards. The challenge to the larger system is to support these policies with necessary resources. (NRC 1996, 232–33)

For students who never attained the literacy goals of the *Standards* in high school, college becomes the last hope—there is nowhere else for them to go. Students who are not majoring in the sciences may be in our classrooms for one or two semesters. When we design their science courses we must be acutely aware that we are very possibly providing the last piece of formal science education they will ever have for the rest of their lives. If they fall through the cracks here, there is no next level to rescue them.

Students who have managed to complete high school without attaining scientific literacy often have special needs. With these students, the instructor, even if he or she is to have merely adequate outcomes, must rely on the full power of the science education system. On the campus level, comprehensive instructional teams composed of reading, writing, math, information systems, tutoring, and academic counseling specialists need to focus on meeting the developmental needs of the struggling students, regardless of the source of their struggles. Along with the college system, private industry, government agencies, and professional societies (i.e., the entire science education system) must come to recognize the importance of these students as a powerful electoral constituency, consumers of science products, and future parents who will be the very first science teachers of our next generation of students. The *National Science Education Standards* ask all components of our national science education system to recognize, value, and support the many levels of students sitting in our science classrooms and laboratories.

## Policy Review for Unintended Effects

All policy instruments must be reviewed for possible unintended effects on the classroom practice of science education.

*Source:* National Research Council. (1966) *National Science Education Standards.* Washington, DC: National Academy Press, 233.

This Standard makes a powerful value statement: "For schools to meet the *Standards*, student learning must be viewed as the primary purpose of schooling, and policies must support that purpose" (NRC 1996, 233).

If we put student learning at the center of purpose, we must put teaching right there beside it—teaching and learning are inseparable. Many colleges and universities have weighted their reward systems in the direction of research and publication. One can argue strongly for a synergy between research and teaching, but in many cases the balance has been skewed to undervalue teaching excellence in promotion and tenure decisions. System Standard F supports the scholarship of teaching. The

Coalition for Education in the Life Sciences (CELS) and the Society for College Science Teachers (SCST) have produced documents that further support the scholarship of teaching (CELS 1998; SCST 1998). Standard F also calls for the review, by "those who actually implement science education policies," of "[a]ll policy instruments...for possible unintended effects on the classroom practice of science education" (NRC 1996, 233). The faculty governance systems in place at most colleges and universities allow for the sort of review called for by this Standard. For other subsystems of science education—government, the private sector, and national societies—science educators need to be cognizant of the avenues for review that exist with these organizations, must be aware of their emerging policy instruments, and must seek opportunities to participate in their review.

It is significant that System Standard F speaks to the review, and thus the valuation, of policies relative to their "effects on the *classroom practice* [my italics] of science education." The *Standards* place the classroom at the focal point, the place where all the system components come together for the good of the student: the classroom is the *raison d'être* for the entire system. This standard can be a foundation to support more student-centered institutional policies, and it implies that teachers or their representatives ought to be major players in policy decisions that will affect the classroom.

## Individual Responsibility———————————

> Responsible individuals must take the opportunity afforded by the *Standards*-based reform movement to achieve the new vision of science education portrayed in the *Standards*.
>
> *Source:* National Research Council. (1996) *National Science Education Standards.* Washington, DC: National Academy Press, 233.

The *National Science Education Standards* are calling everyone in the business of science education to take every opportunity to promote the vision. That these individuals "must take the opportunity afforded by the *Standards*-based reform movement" to do so needs some explanation.

The implication is that since there is a "*Standards*-based reform movement" underway, the movement itself provides support to individuals who are moving in that direction. The National Research Council (NRC) seems to be suggesting here that the *National Science Education Standards* document and the reform movement it is fueling can work to defend innovative teaching strategies where they are not fully appreciated. To a degree, the momentum that the reform movement is acquiring will make *Standards*-based innovation less risky to the innovator than it used to be.

Embodied in System Standard G, too, is the recognition that teachers will teach as they were taught and that students who are in college today need to be taught in tune with the *Standards* or we risk that they will perpetuate what the *Standards* are trying to correct. Individual college professors have the opportunity to break the teach-as-I-was-taught cycle, which has retained some ineffectual methods (see Chapter 1). If we replace the traditional techniques in that cycle with *Standards*-based methods, we can make an important difference.

System Standard G is the last of the System Standards and the last of all the *Standards* listed by the NRC. It seems appropriate that the last point mentioned—writers know that the last point is the one that remains freshest in the reader's mind—gives special importance to the individual's effort. This is, no doubt, in deference to the incontrovertible fact that—regardless of coordination and communication among all sorts of subsystems and regardless of the many systemic initiatives and statewide frameworks—if the reform movement is going to happen at all, it is going to happen because individual teachers are going to make it happen one classroom at a time.

Can any one teacher, or even a whole academic department, pull together all parts of the science education system, maintain a productive dialogue with all its components, and bring all the power they offer to bear on the classroom? Of course not, but the System Standards establish an ideal and urge us to reach out to one another and join hands to achieve it. The following vignette shows just how powerful teachers can be in catalyzing change in undergraduate education.

## From the Field

### Postsecondary Teachers of Science: Catalysts for Change

Mario W. Caprio, Department of Biology
Volunteer State Community College

Constructivism: a theory of learning in which people construct new knowledge by building on what they already know.

We didn't call it constructivism then, nor did we have the *National Science Education Standards* to guide us, but in 1978 the biology faculty at Suffolk Community College in Selden, New York, developed a nontraditional, second-semester biology course for nonscience majors that would be remarkably well-aligned with national standards that were still nearly two decades away. Was this brilliant insight on our part? Were we on the leading edge of what would become a reform movement in science education? Or, were we merely reacting to something else happening at the time?

Interesting questions, perhaps, but why write about "ancient" history here? I think there are two good reasons for doing so. First, it is significant that this course (BY18—Topics in Human Biology) appeared before the reform movement had any real organization. And, second, the course design and activities had an interdisciplinary flavor and saw students as stakeholders who brought their various intellectual assets to share at the table. Constructivism and interdisciplinary thinking in science courses is clearly the language and spirit of the *National Science Education Standards*, and here they were in 1971, at a community college on eastern Long Island, far from the National Research Council.

While I like to think that designing a nontraditional biology course in the seventies made us unique among science educators, that simply is not true. In the late sixties and early seventies, the antiwar demonstrations and general student unrest called for social change and challenged the conservative colleges and universities to redesign curricula for more real world applications. The word *Relevance!* was shouted across campuses and the nation in a tone that was a mixture of anger and urgency. Eventually the colleges listened, and the mid-seventies became a time of experimentation in science education across the country, especially in the nonmajors courses.

The invention and implementation of innovative courses may have been in response to social issues, but that pressure was just the activation energy; we—the teachers—sustained the reaction. We believed in what we were doing. We were enthusiastic. Our students were learning. And, our admin-

istrators generally understood that the new approaches were steps into the future. The point is that long before the *National Science Education Standards* appeared, colleges and universities were implementing pieces of what the *Standards* would be. It is true that many of us are still mired in traditionalism, but it is also true that nearly every postsecondary institution has taken some significant, innovative strides in science teaching and programs. Today, when we hear defenses for traditionalism in science teaching, they generally take the form of the defendants explaining why the constraints under which they must work (e.g., research responsibilities, class size, teaching load) prevent them from implementing a constructivist approach; it is rare to hear anyone argue that traditionalism in science teaching is better teaching.

Noting that innovative teaching in colleges has been, for many years, in line with what the *Standards* now promote strongly suggests that colleges are already *Standards*-based in spirit if not always in deed. But there is another, perhaps more compelling, piece of evidence for their commitment to this ideal: graduate school.

In graduate school, learning is by inquiry: Students collaborate with one another and with their instructors; they learn by example, do projects, and work independently. Yes, colleges and universities have been practicing *Standards*-based teaching for quite a while.

Now, let us go back for a closer look at BY18. How did it become student-centered, incorporate real world is-sues, introduce interdisciplinary per-spectives, and integrate biology into the student's knowledge base? There is only space here for a brief sketch.

On the first of day of class, the BY18 instructor and students worked to de-velop the course's topical outline. That spoke to the outcries for rele-vance and demands for a more signifi-cant student involvement in educa-tional decision-making, and it also created a shared responsibility for the quality of the course. Since the stu-dents were not science majors, they had more knowledge in other disci-plines and were—for the purposes of this course—considered "experts" in those fields. So, in discussions of is-sues (a change in government regula-tions affecting the shrimping industry, for example), the students played their roles as businesspersons, teachers, so-cial workers, journalists, and enforce-ment officers. The instructor was the resident biologist in these conversa-tions. This helped students under-stand the relevant biology and how it interacted with their disciplines. For group projects, the instructor grouped the students according to their major fields and directed that each project involve both biology and the students' majors. The instructor was the biolo-gist member of every group, coaching in the group's project preparation and participating in its presentation. More than just interdisciplinary for the sake of enrichment, the thought here was that if we could link biology to stu-dents' majors, there would be a better chance of them carrying the science with them beyond their BY18 course.

Constructivism, collaborative learn-ing, interdisciplinary thinking, and all the other "innovative" ideas we see in the *National Science Education Standards* are not really new to colleges and uni-versities. We may have retreated to cost-effective traditionalism, but we know a better way to teach. The *Standards* are leading us back to those ideals.

## References

Boyer, E. L. (1990) *Scholarship Reconsidered*. San Francisco: Jossey-Bass.

Coalition for Education in the Life Sciences (CELS). (1998) *Professional Societies and the Faculty Scholar: Promoting Scholarship and Learning in the Life Sciences*. Madison, WI: CELS.

Cuban, L. (1999) *How Scholars Trumped Teachers*. New York: Teachers College Press.

Middleton, S. (1997) The Scholarship of Teaching. *Teaching and Learning in Higher Education* (Dec.).

National Research Council (NRC). (1996) *National Science Education Standards*. Washington, DC: National Academy Press.

Rutledge, J. (1996) Shaping the Future: A National Working Conference (NSF 97-7 1996), *DUE News 1996*. Arlington, VA: National Science Foundation.

Schamel, D., and Gordon, L. (1997) Cost-effective Biology for Elementary Education Majors. In *From Traditional Approaches toward Innovation*, edited by M. W. Caprio. Monograph Series. Greensboro, SC: Society for College Science Teachers.

Society for College Science Teachers (SCST) (1998) Position Statement on the Scholarship of College Science Teaching. *Journal of College Science Teaching* (Sept/Oct.).

University of California (UC). (1997) National Science Education Standards: Impact on Postsecondary Education, an all-University of California Conference. Conference Proceedings, edited by M. Caserio.

# Making Science Education Accessible to All

William H. Leonard

he primary thrust of *Science for All Americans* (AAAS 1989) was that a full and functional science education through grade twelve was an expectation for all persons in the United States. The *Benchmarks for Science Literacy* (AAAS 1993) that followed and, later, the *National Science Education Standards* (NRC 1996) both made recommendations for the specific understandings that should be attained in science at various K–12 grade levels. The implications of these standards for higher education have been discussed throughout this book. The latter two documents, although developed by independent organizations but through funding from the National Science Foundation, reiterated that a science education was not only expected through high school, but that it was a necessity for personal survival in an increasingly technological society. Moreover, having U.S. citizens literate in science was a crucial factor in maintaining world economic competitiveness as well as national security.

Although generally considered ideal and ambitious by the science education community, the national standards (collectively small "s") have become the reference point from which K–12 state science frameworks have been and are currently being developed. These frameworks represent each of the states' interpretation of how the national standards are to be implemented, again for *all* students in school.

The vision of science education for all students presents some very real challenges to teachers in classrooms at all grade levels. Today's classrooms include a diverse group of students who learn best in different ways. This diversity of student-learning needs in the general classroom challenges teachers to reexamine their existing curriculums and instructional methods.

One factor influencing the diversity of learning styles is that increasingly larger populations of students with disabilities are being included in regular academic classrooms. The most recent placement data for students with disabilities as reported in the U.S. Department of Education's Eighteenth Annual Report to Congress (1996) tell us that of the 12 percent of elementary and secondary students receiving special education services, 95 percent are served in regular school buildings. Further, the percentage of students served in the academic classroom has continued to grow over the past several years, establishing inclusion as a major trend. Typically, the students who spend the majority of their time in school in the academic classroom are those students whose disabilities are categorized as within the range of mild to moderate. These may include students with learning disabilities, orthopedic impairments, speech and language impairments, and mild emotional disorders.

The largest categorical group in special education that receives instruction primarily in academic classrooms is students with learning disabilities. A learning disability, as defined in the Individuals with Disabilities Education Act (IDEA) (1997), is a disorder in one or more of the basic psychological processes involved in understanding or in using language, spoken or written, which disorder may manifest itself in imperfect ability to listen, think, speak, read, write, spell, or do mathematical calculations.

Typically, students with learning disabilities have average to above-average intelligence, but because of the disability may have difficulty when faced with traditional academic tasks. The IDEA definition includes conditions such as "perceptual disabilities, brain injury, minimal brain dysfunction, dyslexia, and developmental aphasia." However, it does not include those learning problems that are caused by visual, hearing, or motor disabilities; mental retardation; emotional disturbance; or environmental, cultural, or economic disadvantage.

The inclusion of special education students in general academic classrooms began in elementary schools and now occurs also in secondary schools. Although students with physical disabilities have been a part of the scene in higher education for decades, students with identified learning disabilities are now appearing there in increasing numbers. The commitment to move toward meeting national and local standards may leave many general education teachers frustrated because of a lack of training in, and institutional support for, providing for the educational needs of students with disabilities (Houck and Rogers 1994). At the same time the national reform movement has also caused teachers to feel uncertain about their ability to teach some science concepts (Anderson and Fetters 1996).

It has been suggested that the difficulty in teaching students with disabilities in academic courses at the secondary and postsecondary levels may be due to a number of factors, including the following: (1) lack of monitoring of progress in the general class by a person with knowledge of special education; (2) lack of teacher training in the education of students with disabilities; (3) college students with disabilities lacking basic academic skills and the strategies needed for success; (4) university faculty trained as content specialists who are unwilling or unable to make adaptations and accommodations for students with disabilities who are having difficulty with course content; and (5) students with learning disabilities being discouraged from majoring in more difficult majors, such as the natural sciences. The increasing challenge for college faculty is to provide a variety of means to accommodate the learning of students with learning disabilities. Agencies such as the National Science Foundation feel that providing accommodations for capable students with learning disabilities in a college or university setting is well worth the effort in terms of the valuable contributions they will make to themselves, their profession, and society.

It is also becoming clear from the research that even students who are not classified with learning disabilities learn most effectively in different ways (Leonard 1997). The term *learning styles* has been used extensively in the literature to describe the ways in which individual students learn best. For example, individuals have a preference for learning through one or more of the different senses (Jung 1970). Concrete learners rely more on touch, taste, and smell, and more intuitive and abstract learners prefer hearing and sight. Meyers and McAulley (1958) believe that learning style preferences are related to personality type. For example, sensory learners depend on experiences taken in through their senses; intuitive learners benefit from discussions of abstractions; feeling learners tend to relate what they learn to their own personal and/or societal values; thinking learners benefit most from a logical progression of organized and related concepts. Krause (1996) believes that intuitive-feeling learners are the most endangered when taught by traditional American schooling methods such as lecture.

There is much debate in the literature on learning styles, primarily because of the different ways in which learning styles are categorized. Some evidence indicates that learning is most effective if a student is given information about his or her unique learning style preference and then is provided instruction that takes into account that particular style (Krause 1996). It has been suggested that most instructors teach using their own preferred learning style and ignore the fact that most of the students in their class learn better in other ways. Given that there do seem to be style preferences in the way individuals learn, instructors may be well advised to try to diversify their teaching methods to accommodate the learning needs of a diverse student population.

*There is good news!* The more active learning you can provide your students, the more they will all learn (Leonard 1989). Active learning requires instructional methods that engage students in processing and developing their own understandings of what is presented to them (Caprio 1994; Cannon In press). The key to active learning is allowing the student to interact with the instructor, data, and materials during the learning process. Active learning is especially important for students with learning disabilities and with learning styles different from that of the instructor because active learning allows students to use many senses and many modes or styles of learning. Laboratory instruction, the most common form of active learning in science, accommodates most learning disabilities and most learning styles (Leonard 1997). However, laboratory learning needs to give students opportunities to *think* about what they are doing; this means moving away from a "cookbook" approach to laboratory experiences. The real challenge of implementing active learning in higher education instruction lies with the lecture component of college and university courses. The authors of this book hope that the many suggestions presented here will help you to make all of your instruction more active—and therefore more interesting and productive—for your students.

## References

American Association for the Advancement of Science (AAAS). (1989) *Science for All Americans.* New York: Oxford University.

American Association for the Advancement of Science (AAAS). (1993) *Benchmarks for Science Literacy.* New York: Oxford University.

Anderson, C. W., and Fetters, M. K. (1996). Response: Science Education Trends and Special Education. In *Curriculum Trends, Special Education, and Reform: Refocusing the Conversation*, Special Education Series, edited by M. C. Pugach and C. L. Warger. New York: Teachers College Press.

Cannon, J. (In press) Further Validation of the Constructivist Learning Environment in College Science Courses. *Journal of College Science Teaching.*

Caprio, M .W. (1994) Easing into Constructivism. *Journal of College Science Teaching* 23, 210-12.

Houck, C. K., and Rogers, C. J. (1994) The Special/General Education Integration Initiative for Students with Specific Learning Disabilities: A Snapshot of Program Change. *Journal of Learning Disabilities* 27, 435-53.

Jung, C. G. (1936/1970). *Analytical Psychology, Its Theory and Practice.* New York: Vintage Books Krause, L. B. (1996) *An Investigation of Learning Styles in General Chemistry Students.* Ph.D. diss., Clemson University.

Leonard, W. H. (1989) Ten Years of Research on Science Laboratory Instruction at the College Level. *Journal of College Science Teaching* 18, 303-306.

Leonard, W. H. (1997) How Do College Students Learn Science? In *Methods of Effective Teaching and Course Management for University and College Science Teachers,* edited by E. D. Siebert, M. W. Caprio, and C. M. Lyda. Dubuque, IA: Kendall/Hunt.

Meyers, I., and McAulley, M. (1958) *Manual: A Guide to the Development and Use of the Myers-Briggs Type Indictor.* Palo Alto, CA: Consulting Psychologists Press.

National Research Council (NRC). (1996) *National Science Education Standards.* Washington, DC: National Academy Press.

U.S. Department of Education. (1996) *Eighteenth Annual Report to Congress.* Washington, DC: U.S. Department of Education.

# Profiles of Contributors

**Mario W. Caprio** retired from Suffolk Community College in 1996 after teaching there for thirty-three years. In 1998, he received the rank of Professor Emeritus from that school and now teaches biology and integrated science courses at Volunteer State Community College. In addition to his teaching responsibilities, Professor Caprio is on the Board of Directors of the Biological Sciences Curriculum Study and the BioQUEST Curriculum Consortium and is part of the Editorial Advisory Board of the *Journal of College Science Teaching (JCST)*. Professor Caprio is a charter member of the Society for College Science Teachers and currently serves as that organization's secretary/treasurer. He frequently makes presentations about the reform movement in science education at National Science Teachers Association meetings and writes and edits articles about community college issues for his regular column ("The Two-Year College") in the *JCST*.   e-mail: *mcaprio@mwsi.net*

**Brian P. Coppola** is an associate professor of chemistry and the coordinator for undergraduate organic chemistry curriculum at the University of Michigan–Ann Arbor and a faculty associate at the University of Michigan Center for Research on Learning and Teaching. He received his B.S. in 1978 from the University of New Hampshire and his Ph.D. in organic chemistry from the University of Wisconsin–Madison in 1984, having joined the faculty at the University of Wisconsin–Whitewater as an assistant professor in 1982. In 1986, after moving to Ann Arbor, Dr. Coppola joined an active group of faculty in the design and implementation of a revised undergraduate chemistry curriculum. His recent publications range from mechanistic organic chemistry research in 1,3 dipolar cycloadditions, to educational philosophy, practice, and assessment. Dr. Coppola is a Pew Scholar affiliated with The Carnegie Foundation for the Advancement of Teaching's CASTL program.   e-mail: *bcoppola@umich.edu*

**Marvin Druger** is a professor of biology and science education and chair of the Department of Science Teaching at Syracuse University. He has a Ph.D. in zoology (genetics) from Columbia University and has taught introductory college biology to many thousands of students over a span of more than forty years. He has served as president of three international science education organizations: the Society for College

Science Teachers (SCST); the Association for the Education of Teachers in Science (AETS); and the National Science Teachers Association (NSTA). Dr. Druger also served as chair and secretary of the education section of the American Association for the Advancement of Science (AAAS) and program director for the Science and Mathematics Education Networks Program at the National Science Foundation. He is a fellow of the New York Academy of Sciences, a fellow of AAAS, and currently president-elect of SCST. Professor Druger has received several science teaching awards, including the Meredith Professorship for Teaching Excellence at Syracuse University. e-mail: *druger@SUED.SYR.edu*

**Suzanne S. Drummer** is the material evaluation coordinator at the Eisenhower National Clearinghouse for Mathematics and Science Education in Columbus, Ohio. She has sixteen years experience teaching high school science and five years teaching undergraduate science. She earned an M.S. in zoology from the University of Kentucky and is currently working on a Ph.D. in science education at The Ohio State University. e-mail: *sdrummer@enc.org*

**Diane Ebert-May** (Ph.D., University of Colorado–Boulder) is the director of Lyman Briggs School, a residential liberal arts science program within the College of Natural Sciences at Michigan State University (MSU) and is a professor in the MSU Department of Botany and Plant Pathology. Her research, currently funded by the National Science Foundation, focuses on alternative assessments for large science classes and variables that affect long-term faculty change. She is a member of the National Research Council Committee on Recognizing, Evaluating, and Rewarding Excellence in Undergraduate Teaching in Science, Mathematics, Engineering, and Technology and is on the board of directors of the National Association of Research in Science Teaching. Her ecological research continues on Niwot Ridge, Colorado, where she has conducted long-term research on alpine tundra plant communities since 1971. e-mail: *ebertmay@pilot.msu.edu*

**Patrick Gleeson** is a professor in the Department of Physics and Pre-Engineering at Delaware State University (DSU). He received his Ph.D. in physics from the University of Delaware and has been with the physics department at DSU since 1972. Since coming to DSU, he has been involved in condensed matter studies and, more recently, medical physics investigations. Dr. Gleeson supervises the physics education program, but his primary responsibility is teaching. He teaches physics and some mechanical engineering courses at levels ranging from conceptual courses for nonmajors through graduate course work in physics.   e-mail: *pgleeson@dsc.edu*

**Ben Golden** is a professor emeritus of biology at Kennesaw State University (KSU). He has a long record of innovation in both teaching and curriculum design. He started designing interactive computer programs in the early 1980s and has more recently designed interactive, computer-graded lab reports that give students instant

feedback. Professor Golden was one of the earliest faculty members at KSU to adopt student-centered learning activities and was instrumental in making such techniques the central pedagogy in the integrated science sequence at KSU. He has been a Georgia Science Teachers Association College Teacher of the Year.
e-mail: *bgolden@kennesaw.edu*

**Theodore D. Goldfarb** is a professor of chemistry at the State University of New York–Stony Brook. His initial scholarly interests were in the fields of molecular spectroscopy and dynamics. In the late 1970s he began a transition that has culminated in his present research, teaching, and publication focus on environmental problems and issues of science and public policy. Dr. Goldfarb's teaching related to ethical issues in the sciences includes undergraduate courses and seminars, ethics components of a graduate course entitled "Controversial Issues in the Environmental Sciences," and Chautauqua courses for college teachers. He has given invited lectures at Western Michigan University's Center for the Study of Ethics in Society and served as the facilitator of the primary and secondary education workshop at the conference, "Biotechnology and Ethics: Blueprint for the Future," held at the Northwestern University Center for Biotechnology. Dr. Goldfarb is editor of the book *Taking Sides: Clashing Views On Controversial Environmental Issues,* now in its eighth edition.
e-mail: *tgoldfarb@notes.cc.sunysb.edu*

**Leslie Sears Gordon** (Ph.D., University of Idaho) worked as a teacher of gifted elementary students for over twenty years before returning to school to get her doctorate in 1995. She received the Presidential Award for Excellence in Math and Science Education and the Milken National Educator Award. She is also listed in *Who's Who Among America's Teachers*. Dr. Gordon has taught science and math courses for teachers and student teachers for over a decade. She is now working on three National Science Foundation projects in Alaska. The latest project is a summer institute for teachers that integrates Western science, Native knowledge, and best practices in math and science education.    e-mail: *lgordon@northstar.k12.ak.us*

**Lauralee Guilbault** is an assistant professor of chemistry at Alverno College. She earned her Ph.D. from the University of Tennessee. She is a member of the Valuing Ability Department and the special committee on intermediate students. Dr. Guilbault is co-coordinator of the environmental science major and has been the recipient of a chemistry instrumentation grant. She is a member of the Goals 2000 consortium with the Milwaukee Public Schools and eleven higher education institutions of Wisconsin. She is vice president of Women in Science of Southeastern Wisconsin and works with Science for Children programs throughout the year.
e-mail: *lauralee.guilbault@alverno.edu*

**George Gurria** is a retired associate professor of chemistry at Alverno College, where he was a longtime member of the Analysis Ability Department and faculty of the annual Workshop for College Educators. He earned his Ph.D. from Johns Hopkins University. He was chief health profession advisor at Alverno and served as the dean of the Natural Sciences Division and consultant to colleges in the United States and Canada on outcome/ability-based education, critical thinking, and assessment.

**Robert W. Harrill** serves as executive director of the Institute for Conservation Studies (ICONS) at the College of Santa Fe and as an associate professor in the college's new major program in conservation science. He is founder of EarthViews, an organization that specializes in remote sensing applications in conservation and natural resource management. Dr. Harrill was executive director of LightHawk, a conservation organization that uses the power of flight in its efforts to save threatened ecosystems, and he was senior associate at the Woods Hole Research Center. Dr. Harrill has developed college-level environmental science programs and has broad administrative experience in nonprofit organizations. He was a faculty member at the University of California–Los Angeles (UCLA), New College (now a part of the University of South Florida), and Prescott College and was co-director of the environmental science program at Prescott. Dr. Harrill served as academic vice president and acting president at Prescott College and as vice president of institutional advancement at Bryant College. He graduated from Grinnell College and earned his doctorate at UCLA.    e-mail: *icons@csf.edu*

**Janan M. Hayes** is a professor of chemistry and physical science and vice president of instruction at Merced College. She was formerly dean of science at Cosumnes River College and a faculty member at American River College. Dr. Hayes is co-principal investigator of Project Inclusion and has served as project director of various efforts to improve the education and training of science-math teachers and of correctional officers in the California Department of Corrections. She is a member of the Council of the American Chemical Society (with numerous national governance committee assignments), the Two-Year College Chemistry Committee, and the California Association of Chemistry Teachers. She has made a number of presentations at local, regional, and national meetings of these organizations and organized and presided over several symposia. Dr. Hayes has also served on various National Science Foundation review panels. She earned a Ph.D. in chemistry from Brigham Young University in 1971.    e-mail: *jmhayes@merced.cc.ca.us*

**Elizabeth T. Hays**, Ph.D., is an associate professor of physiology at Barry University. She holds appointments in both the School of Natural and Health Sciences and the School of Graduate Medical Sciences, where she teaches medical physiology for the podiatric medical students and physiology at various postsecondary levels; she mentors undergraduate students in research projects in muscle regeneration and physiology. Dr. Hays designed, developed, and implements a program for at-risk freshman biology majors, a component of Barry University's Minority Access to Research Careers (MARC-USTAR) grant. She is a life member of the Florida Association of Science Teachers, serves on the Board of Directors of the National Science Teachers Association as director of the multicultural division, is a former councilor-at-large for the Society of College Science Teachers, is active in the Association for Multicultural Science Educators and the American Physiological Society, and is councilor-at-large for the Florida Academy of Sciences.    e-mail: *ehays@mail.barry.edu*

**Judith E. Heady** is an associate professor of biology at the University of Michigan–Dearborn. She received her B.A. in biology from Cornell College, an M.S. in zoology from the University of Iowa, and a Ph.D. in developmental biology from the University of Colorado–Boulder. Her research has shifted from control of early embryonic development in amphibians to learning in the undergraduate science classroom. Dr. Heady has been a councilor-at-large on the executive board of the Society for College Science Teachers (SCST) and an editor of the annual SCST program and abstracts. She has presented papers on her introductory and upper-level biology classes in the SCST sessions at the regional and national meetings of the National Science Teachers Association, the American Association for Higher Education Assessment Conference, and the Lilly-Atlantic meeting of the International Alliance of Teacher-Scholars. In fall 1995, she was a distinguished visiting professor of biology at the University of Wisconsin–La Crosse as part of a National Science Foundation-sponsored program to improve introductory science teaching.    e-mail: *jheady@umich.edu*

**Gordon P. Johnson** completed his B.A. in the natural sciences and mathematics at Augsburg College, his M.S. degree at the University of Wisconsin–Madison, and M.A. and Ph.D. at the University of Minnesota–Minneapolis. His teaching experience includes thirteen years at the middle and secondary levels in public schools in Minnesota, followed by thirty years in the Department of Physics and Astronomy at Northern Arizona University. Dr. Johnson's responsibilities at the university included the designing and teaching of science content based on teacher education courses as well as the teaching of the general physics course for nonmajors. Dr. Johnson is the recipient of the Distinguished Faculty Award and the Teaching Scholar Award from Northern Arizona University and the Distinguished Service to Science Education Award from the National Science Teachers Association.
e-mail: *Johnson@bohr.phy.nau.edu*

**Gerald H. Krockover** (Ph.D., University of Iowa) is a professor of Earth and atmospheric science education at Purdue University. He holds a joint appointment between the Department of Earth and Atmospheric Sciences in the School of Science and the Department of Curriculum and Instruction in the School of Education. He conducts programs to prepare graduate teaching assistants in science and has received the Purdue University Outstanding Undergraduate Teaching Award, the Purdue University Impact on Learners Faculty Teaching Award, the National Association of Teacher Educators Distinguished Major Professor Award, and the Association for the Education of Teachers in Science Outstanding Science Educator Award. He is cited

in *American Men and Women in Science*, *Who's Who in Science and Engineering*, and *Five Hundred Leaders of Influence* and has received more than $2.2 million in external funding to improve science education throughout the United States.
e-mail: *hawk1@purdue.edu*

**William H. Leonard** (Ph.D., University of California–Berkeley) is a professor of science education and biology at Clemson University. His research activities involve inquiry learning in science, science laboratory instruction, reading science text, and learning science through computer technologies. He coordinates seminars on inquiry instruction for Clemson University science, math, and education faculty in hopes that faculty might model national standards teaching strategies for future teachers of K–12 science and mathematics. An active biology curriculum developer, he is an author of *Laboratory Investigations in Biology, Biological Science: An Ecological Approach,* and *Biology: A Community Context*. He has won university awards for both teaching and research. e-mail: *leonard@mail.clemson.educ*

**Gary Lewis** is a professor and the director of digital instructional development at Kennesaw State University. He has a Ph.D. in physics from Georgia Institute of Technology and has worked in both industry and academics, with current interests in electronics, optics, general science, and the use of technology in teaching. His background includes a variety of engineering and science disciplines, from nuclear physics and solar energy to solid-state physics and integrated circuit manufacturing. e-mail: *glewis@kennesaw.edu*

**William J. McIntosh** (Ph.D., Temple University) is a professor of science education at Delaware State University, where he teaches courses in introductory physical and Earth sciences as well as science methods courses. He is interested in the role of college and university faculty in systemic reform; he has pioneered changes in his own science courses and works collaboratively with faculty from other institutions to promote course reform. He is past president of the Society for College Science Teachers and serves on the Board of Directors for the National Science Teachers Association as director of the college division. e-mail:*bmcintsh@udel.edu*

**Susan Millar** (Ph.D., cultural anthropology, Cornell University) directs the University of Wisconsin (UW)–Madison Learning through Evaluation, Adaptation, and Dissemination (LEAD) Center (*www.engr.wisc.edu/~lead*). The LEAD Center, an organization of twelve professionals, supports faculty (primarily in the science and engineering fields) engaged in educational reform activities by providing evaluation research. She also is the "Lead Fellow" for the Institute on Learning Technology, a project of the National Institute for Science Education College Level One team (*www.wcer.wisc.edu/nise/cl1*). Dr. Millar participates on the national advisory boards of various organizations seeking to improve science learning in higher education, including the Advisory Board for the Education and Human Resources Directorate of the National Science Foundation. Previous positions include lecturer for the UW–Madison Women's Studies Program, policy analyst for the UW System Administration, co-director of the National Study of Master's Degrees, and research associate for the Penn State Center for the Study of Higher Education. e-mail: *smillar@engr.wisc.edu*

**Kent S. Murray** is an associate professor of geology and environmental science at the University of Michigan–Dearborn where he teaches undergraduate and graduate courses in groundwater hydrology, watershed analysis, and environmental geology. He received his Ph.D. in geology at the University of California–Davis and has worked for the U.S. Geological Survey and the California Energy Commission as a geothermal specialist. He currently serves as a consultant to numerous national and international environmental petroleum and environmental law firms. His research interests, which have been supported with funding from the National Science Foundation and the Environmental Protection Agency, include contaminant hydrology, geothermal resource assessment, the relationship between groundwater and surface water quality, and the impact of urban watersheds on the water quality of the Great Lakes. He has become increasingly interested in undergraduate education in the geosciences and is a contributing author to the *Journal of Geoscience Education*.
e-mail: *kmurray@umich.edu*

**Jeanne L. Narum**, director, Independent Colleges Office, became the founding director of Project Kaleidoscope (PKAL) in 1989. She continues to develop and coordinate the various facets of PKAL, including the Faculty for the 21ˢᵗ Century Network, seminars and publications on facilities planning, and workshops and events on disciplinary, topical, and institutional issues (increasingly, issues relating to technologies in undergraduate science, mathematics, and engineering are being addressed in PKAL workshops). Ms. Narum was publisher for PKAL Volume I, and editor-in-chief for Volumes II and III. She participated in the 1998 Aspen Institute for Higher Education. She is a member of the Board of Trustees at Lenoir-Rhyne College, a member of the National Research Council Committee on Recognizing, Evaluating, and Rewarding Excellence in Undergraduate Teaching, and a councilor in the Council for Undergraduate Research. She holds an honorary doctorate from the University of Portland. e-mail: *pkal@pkal.org*

**Susan K. Painter**, is manager of academic programs in the Department of External Scientific Affairs at Merck Research Laboratories, a division of Merck & Co., Inc. Ms. Painter holds a B.S. degree in animal science from the University of Tennessee, Knoxville; she has fifteen years of research experience in biochemistry—encompassing academia, biotechnology, and the pharmaceutical industry. She is a member of the Volunteer Coordinating Board of the Merck Institute for Science Education, which supports science education initiatives in target school districts to enhance and improve science, math, and technology education through hands-on exploration. Fostering industry support of undergraduate science research is one of Ms. Painter's primary

responsibilities through the Merck/AAAS Undergraduate Science Research Program, a competitive award program to promote the interdisciplinary relationship between chemistry and biology. In addition, Ms. Painter helps to identify related avenues for Merck undergraduate support via established external programs including the Council on Undergraduate Research's Summer Fellowship Program and the University of Minnesota's Summer Undergraduate Research Program.
e-mail: *susan_painter@merck.com*

**Robert A. Paoletti** is a professor of biology at King's College. His research interests include signal transduction and regulation of gene expression, especially in developing organisms; he uses cell culture model systems in student laboratories to provide experience for students in these and other areas of current importance. Dr. Paoletti has been involved in the King's College course-embedded model of assessment in higher education and has, for a number of years, delivered relevant papers at the annual meetings of the American Association for Higher Education. He has also consulted on assessment in the sciences at colleges and universities nationwide.
e-mail: *rapaolet@gw02.kings.edu*

**Patricia L. Perez** has been a professor of chemistry at Mt. San Antonio College, a public, two-year community college, since 1968, serving as department chairperson from 1989 to 1993. She is co-principal investigator of Project Inclusion, a National Science Foundation-sponsored effort to focus student attention on the scientific contributions of members of various underrepresented groups. In addition, she is a consortium participant in the Molecular Science Education Project, a University of California-Los Angeles–California State University-Fullerton college alliance for systematic curriculum reform. Professor Perez is a member of the American Chemical Society, the Two-Year College Chemistry Committee, and the California Association of Chemistry Teachers. She has made numerous presentations at local, regional, and national meetings of these organizations and organized and presided over several symposia. Additionally, she has served on various National Science Foundation review panels and workshops. She earned an M.S. in chemistry from UCLA in 1968.
e-mail: *pperez@mtsac.edu*

**Stanley H. Pine** is a professor of chemistry at California State University–Los Angeles, and past chair of the American Chemical Society (ACS) Committee on Education. He continues service on that and several other education and laboratory safety related groups. He served as a program director in the National Science Foundation Division of Undergraduate Education and is a member of the ACS College Chemistry Consultant service that provides guidance to two-year and four-year chemistry departments. Dr. Pine earned his Ph.D. degree at UCLA and continued postdoctoral work at Harvard. He has been a Fulbright Scholar, a visiting professor at several universities, and a recipient of continuous research support from agencies such as the National Science Foundation, the National Institutes of Health, and the Petroleum Research Fund. e-mail: *spine@calstatela.edu*

**Barbara Woodworth Saigo**, biologist, is president of Saiwood Biology Resources and Saiwood Publications. She has been a university biology faculty member, director of a university grants and contracts office, textbook author (environmental science, botany, biology, professional development), grant project director, and national grant proposal reviewer. She serves as a consultant and advisor to K–12 schools, universities, and state agencies for science education, staff development, manuscript and publication development, grant writing, and integration of instructional technology. Her recent research focus is on teacher implementation of research-based instructional strategies and its impact on student learning in a content area, and on the role of conceptual change in both learning and group processes. e-mail: *bsaigo@aol.com*

**Douglas Schamel** (M.S., wildlife management, University of Alaska–Fairbanks; Ph.D. candidate in behavioural ecology, Simon Fraser University) is an assistant professor of biology at the University of Alaska–Fairbanks, where he has served as a faculty member for twenty years. He teaches introductory biology for majors and nonmajors, an upper-division ornithology course, and science for elementary teachers. His research focuses on the behavioral ecology of nesting shorebirds in northern Alaska, especially phalaropes (sea-going shorebirds). His true passion, however, is science education and outreach. Mr. Schamel has served on several local and regional advisory boards for science education, has coordinated the Fairbanks area K–8 science fair for a decade, co-founded the Alaska Regional Junior Science and Humanities symposium, and is opening an interactive science center in Fairbanks. He is the recipient of local and national awards for teaching and science outreach, including the Society for College Science Teachers/Kendall-Hunt Outstanding Undergraduate Science Teacher of the Year award. e-mail: *ffdls@uaf.edu*

**Gail Schiffer** is an associate professor of biology at Kennesaw State University. Having participated in the development of the nonmajor science sequence at KSU, she now serves as the general education science coordinator. Dr. Schiffer, along with Dr. Ben Golden, Dr. Diane Willey, and Dr. Gary Lewis, has made numerous presentations at National Science Teachers Association and Society for College Science Teachers conferences on both the design and establishment of the integrated science course as well as the design and validation of the SCIPROS test, a test of science process skills. e-mail: *gschiffe@kennesaw.edu*

**David Seybert** is a professor of chemistry and biochemistry at Duquesne University. He received his Ph.D. from Cornell University and is currently serving as interim dean of the Bayer School of Natural and Environmental Sciences at Duquesne. He also directs the American Association of Colleges and Universities–National Science Foundation's Preparing Future Faculty Program in the Department of Chemistry and Biochemistry. His research interests focus on mechanistic and clinical aspects of lipid peroxidation and the mechanisms by which oxidative by-products damage proteins and other biomolecules. e-mail: *seybert@duq.edu*

**Maureen Shiflett** received her B.A. in biology from Eastern Washington University, her M.S. in food microbiology from Oregon State University, and her Ph.D. in microbiology from the University of Rochester. She was a postdoctoral fellow at Scripps Clinic and Research Foundation, La Jolla, California, and at the University of California–Los Angeles. After several years of educating medical students she became interested in K–12 science education and joined the staff of the team that developed the *National Science Education Standards*. She is co-executive director of the Southern California Regional Office of The National Faculty. She also maintains an active consulting service in standards-based science and mathematics education, partnership building, professional development, and university/school relationships. Dr. Schiflett is the author or co-author of over thirty publications on topics ranging from research microbiology to science education. e-mail: *info@TNF.org*

**Harry L. Shipman** teaches physics, astronomy, and other subjects at the University of Delaware. His primary professional background is in astronomy, and he has published over 150 astronomical articles in journals, as well as four popular books. He also does research in science education and has published in the *Journal of College Science Teaching*, the *Journal of Research in Science Education,* and *Science and Education.* e-mail: *harrys@udel.edu*

**Eleanor D. Siebert** (Ph.D., University of California–Los Angeles) chairs the Department of Physical Sciences and Mathematics and is a professor of chemistry at Mount St. Mary's College. She teaches introductory physical science, chemistry, and physics to both science majors and nonmajors and instrumental analysis and thermodynamics at the upper-division level. Her research activities with undergraduates involve studies of phase separation in model biological systems. She is author of *Experiments for General Chemistry* and *Foundations for General Chemistry*; she has co-edited *Methods of Effective Teaching and Class Management for University and College Teachers of Science*. She is past president of the Society for College Science Teachers and past college director of the National Science Teachers Association. Currently she chairs the Committee on Public Relations and Communications for the American Chemical Society. She is listed in *Who's Who Among America's Teachers* and in *American Men and Women in Science.* e-mail: *esiebert@msmc.la.edu*

**Nannette Smith**, director of the Division of Natural, Behavioral, and Social Sciences at Bennett College, earned her B.S. and M.S. degrees in microbiology and botany from Howard University and her Ph.D. in plant pathology from North Carolina State University (NCSU). She was the first African-American woman to earn a Ph.D. from NCSU. During 2000, Dr. Smith served as the president of the North Carolina Science Teachers Association. e-mail: *nsmith@bennett.edu*

**Joseph I. Stepans**, a professor of science and mathematics education at the University of Wyoming, earned a bachelor's degree in physics from California State University–Stanislaus, and a master's degree in physics and a doctorate in science education from the University of Wyoming. He worked briefly in Iran to develop curricula and teach graduate courses in natural sciences and educational research. At the University of Wyoming he teaches undergraduate and graduate courses in the content and methods of mathematics and science to prospective K–12 teachers. Dr. Stepans's research interests focus on mathematics and science misconceptions and on effective ways to bring about a conceptual change in students' learning. He launched WyTRIAD in 1991, a nationally recognized professional development model that has been implemented in over thirty school districts in six states. In addition, he has authored or coauthored numerous articles and several books including *Targeting Students' Science Misconceptions, Changing the Classroom from Within,* and *Challenging Students to DO Meaningful Mathematics.* e-mail: *jstepans@uwyo.edu*

**Lynda C. Titterington** is the senior science education abstractor for the Eisenhower National Clearinghouse for Mathematics and Science Education in Columbus, Ohio, where she examines the best curriculum materials available and interacts with master teachers from around the nation. She is currently a doctoral candidate in science education at The Ohio State University and teaches nonmajors' biology courses at Columbus State Community College. Many of her students are second-career adults, and she works with them to develop alternative assessments that are more suited to the unique characteristics of adult learners. Her goals are to develop formative assessments to help students reduce anxiety and understand content before final examinations, to find a more authentic audience than the teacher, and to design relevant tasks that help students develop important professional skills in communications and technology. e-mail: *ltitter@blknight.enc.org*

**Stacy Treco** is vice president, director of marketing for Benjamin/Cummings, an imprint of Addison-Wesley Longman Publishers. Ms. Treco joined AWL in 1988 as a sales representative, moved on to be the marketing manager for the life science list, transitioned back into sales as a manager, and then moved back to marketing as director in 1997. Her proudest moments in her publishing career are when she is hosting a *Strategies for Success* workshop or working with one of the gifted authors at Benjamin/Cummings. Ms. Treco graduated with her bachelor's degree from Bucknell University in 1985 and her master's degree from Tulane University in 1987. e-mail: *stacy.treco@awl.com*

**Leona Truchan** (Ph.D., Northwestern University) is a professor of biology at Alverno College where she served as chair of the global perspectives ability department. Currently she is coordinator of the biology department, of the science education program, and of the environmental science major. She has been dean of the Natural Science, Mathematics, and Technology Division, president of the Association of College and University Biology Educators, president of the Association of Biology Laboratory Education, and former board member of the Society for College Science Teachers. Dr. Truchan is a consultant to teachers from K–University on teaching, learning, and assessment in the United States, Canada, Australia, and New Zealand. She has been principal author and co-participant in numerous grants for improving science teaching and assessment, as well as a recipient of multiple Eisenhower grants for elementary teachers. e-mail: *truchal@alverno.edu*

**Diane Willey** (Ph.D., educational psychology, University of Iowa) is a professor of education and director of research and assessment in teacher education at Kennesaw State University. She consults and provides faculty development on teaching and assessing different types of learning outcomes and designing program evaluation. e-mail: *dwilley@kennesaw.edu*

**Jack Winn** is Distinguished Teaching Professor and chair of mathematics at the State University of New York (SUNY)–Farmingdale, where he has taught since 1974. He received his Ph.D. from Adelphi University under the direction of Eugene Levine, his M.S. from C.W. Post College, and his B.S. from SUNY–Oneonta. In 1998 he received the Award for Distinguished College or University Teaching of Mathematics from the Metropolitan New York Section of the Mathematical Association of America (MAA). Dr. Winn is currently chair-elect of the Metropolitan Section of the MAA. His interests include graph theory and curriculum reform in mathematics education. Most recently Dr. Winn has worked on developing SUNY–Farmingdale's major in applied mathematics, an innovative joint program with SUNY–Stony Brook, and he has been very active in the Long Island Consortium for Interconnected Learning. e-mail: *winnJA@SNYFARVA.cc.farmingdale.edu*

**Robert E. Yager** is a professor of science education at the University of Iowa where he also earned two graduate degrees. He has directed over one hundred projects funded by the National Science Foundation and served as chair for nearly one hundred Ph.D. students. Dr. Yager has served as president of seven national organizations, including the National Science Teachers Association and the National Association of Science-Technology-Society (STS). He has been involved internationally with ongoing international projects in Korea, Taiwan, Thailand, Indonesia, and various places in Europe. Dr. Yager's current research interests and teaching are involved with STS as an instructional reform effort. e-mail: *robert_yager@uiowa.edu*

**Dana L. Zeidler** earned his Ph.D. in science education from Syracuse University and is certified in 7–12 science. Presently, he is a doctoral program director for science education at the University of South Florida. He has served as a managing editor for the *Journal of Science Teacher Education* and has been on the editorial board for *Science Education* and the *Journal of Research in Science Teaching*. His research interests have included studies in science education reform goals, students' and teachers' epistemological beliefs in the nature of science, critical thinking and discourse, and ethical reasoning related to socio-scientific issues. He also has been teaching Isshinryu ("One heart way") karate since 1982. e-mail: *zeidler@tempest.coedu.usf.edu*

# Index

## A

Abstract learners, 171
Active learning
  challenge of implementing in higher
    education, 172
  description, xvi
  laboratory instruction and, 172
  learning disabilities and, 171–172
African-Americans
  assessment bias and, 81
  increasing the number of African-American
    scientists, 129
  underrepresentation in science fields, 131
Aging, link with space industry, 20–22
Alaska Science Consortium Learning Cycle
  Model, 153–154
Alternative assessments, 69
Alto Merse Nature Reserve, 122
Alverno College, assessments as student
  learning, 87–89
American Association for Higher Education
  (AAHE), 58, 147, 149
American Association for the Advancement of
  Science (AAAS), 129, xiii
American Association of Colleges of Pharmacy
  (AACP), 130
American Association of Physics Teachers
  (AAPT), 143
American Chemical Society (ACS), 51
  Community Interaction—Student Affiliates
    Grants for Affiliate Intervention, 160
  Division of Chemical Education, 147
  Education Web site, 160
  five areas of graduate-level interest, 148–149
  Office of Education and International
    Activities, 159
  Strength in Numbers: Uniting the Fronts of
    Higher Education symposium, 147–150
  Student Affiliates program, 160
  Task Force on the National Science Education
    Standards, 159
American Foundation for Pharmaceutical
  Education, 131

American Pharmaceutical Association (APhA),
  130, 131
Anne Arundel Community College workshop,
  157
Asbestos exposure, 109–110
Asian-Americans, assessment bias and, 81–82
Asian/Pacific Islanders, underrepresentation in
  science fields, 131
Assessing and Validating the Outcomes of College,
  87
Assessments
  alternative assessments, 69
  bias and, 80–86
  content standard and, 68–69
  course goals and, 61, 75
  courses with large enrollments and, 9–11
  criteria for, 88–90
  critical features of, 60
  curriculum reform and, 78–80
  definition, 57, 58
  evaluation and, 8, 57, 60–61
  examinations, papers, reports, and projects,
    61–62
  goals of, 89
  informal, 69
  interviews, observations, and focus groups,
    63
  large-scale survey work, 63
  link to curriculum content, 60
  link with learning and teaching, 8, 59–60
  literature-based examinations, 84–86
  making sound inferences, 86–90
  matching quality of data to consequences,
    74–80
  most confusing concept question, 10
  multiple assessments, 8–11, 22, 61, 69, 75
  nonscience majors and, 82–84
  opportunity to learn and, 68–74
  peer-based review, 62, 63–65, 71
  performance-based, 62, 65–68
  periodic sampling of intermediate materials,
    62
  pretests and posttests, 76–78

principles of, 58–59
program review and, 118–119
of programs, 117
prompts and, 88
purposes of, 60–61
reaction papers, 106
role of, 57, 59–60
self-assessments, 9, 83
Structured Study Groups, 63–68
student learning and, 87–89
uses for, 57
Association of American Colleges and
    Universities (AAC&U), 50, 51, 148, 160

**B**

Bandelier National Monument, 122
Barry University, Minority Access to Research
    Careers program, 132–133
Basketmaking, dyes for, 113
*Benchmarks for Science Literacy,* 169, xiii
Bias
    African-Americans and, 81
    Asian-Americans and, 81–82
    concept of fairness and, 80
    minimizing, 82
    stereotype threat, 81
    students with physical problems, 82
    subtle and unintentional, 80–81
    women and, 81–82
Brainstorming activities, 105
Budget considerations
    funding agencies, 144–145
    resources to support the science program,
        127–131
    system standards and, 151, 158–159

**C**

California Institute of Technology Precollege
    Science Initiative (CAPSI), 29
Campus Teaching Academy, 148
Canyonlands National Park, 122
Carnegie Academy for the Scholarship of
    Teaching and Learning (CASTL) program,
    134–135, 148, 149–150
Carnegie Foundation for the Advancement of
    Teaching, 59, 134, 147, 148
Carnegie Scholars, 134, 148
CASTL program, 134–135, 148, 149–150
Chemistry. *See* General chemistry
*Chemistry in Context,* 160
*Chemistry in the National Science Education
    Standards,* 159

Coalition for Education in the Life Sciences
    (CELS), 143, 163
College of Santa Fe, conservation science
    program, 120–124
Community Interaction—Student Affiliates
    Grants for Affiliate Intervention, 160
Community of learners concept, 43
Computer-Aided Chemistry (CAChe) software,
    64–68
Concept mapping, 69
Concrete learners, 171
Connecting the First Day workshop, 157
Consciousness-raising exercises, 103
Conservation science program, 120–124
Consistency of science programs, 116–119
Constructivist theory of learning, 21, 101
    description, xv
    differences between traditional and
        constructivist classrooms, xv–xvi
    professional development standards and,
        41–42, 48
    system standards and, 164–166
Content standards
    assessments and, 68–69
    cultural literacy and, 102
    curriculum reform and, 104
    for Earth and space science, 101
    focus of, 96
    "good" science and "bad" science, 107
    history and nature of science, 110–114
    importance of laboratory or field
        experiences, 97–99
    inclusiveness of, 96, 110
    for life science, 100
    nonscience majors and, 101–103
    personal and social perspectives, 107–110
    for physical science, 100
    Project Inclusion, 111–114
    relation between science and technology and
        science in personal and social
        perspectives, 108
    research approach to the general chemistry
        laboratory, 98–99
    science and technology, 104–107
    science as inquiry, 97–99
    society and, 105–107
    subject matter content, 99–103
    unifying concepts and processes, 95
    workshops, 98–99, 113
Coordination of science education policies,
    150–155
Corporations, 142–143
Cost-effective biology for elementary education
    majors, 152–155

Council of Graduate Schools (CGS), 50, 51, 148, 160
Course goals, 61, 75–76
Course sequencing, 120
Creative Teaching Strategies in the Sciences, 157
Cuban-American students, 133
Cultural diversity, 111–114. *See also specific cultural and ethnic groups*
Cultural literacy, 102
Curriculum criteria, 120–124
Curriculum reform, 19, 78–80, 104, 118, 125, 127, 137–138, 158
Cyberspace: Another Dimension to Teaching the Sciences, 157

**D**

Diversity of the student population, 111–114, 169–171
Duquesne University
Center for Teaching Excellence, 52
Shaping the Preparation of Future Science and Mathematics Faculty initiative, 51–53

**E**

Earth and space science content standards, 101
Eisenhower Program, xiii
Elderly persons, 20–22
Environmental issues, 108–110. *See also* Conservation science program
Equilibration concept, 42, 48
Equity
equitable policies, 161–162
in science opportunities, 132–133
Essays, 69
Ethical considerations, course content and, 110
Examinations, papers, reports, and projects, 61–62
"Expanding Your Horizons" conference, 98

**F**

Feeling learners, 171
Field experiences
content standards and, 97–98
program standards and, 122
Formative evaluation, 60–61
Foundations for Excellence in the Chemical Process Industry project, 160
*From Analysis to Action,* 36

**G**

Gabrieleno Tongva people, traditional basketmaking, 113
General chemistry
analysis of an innovative college chemistry course, 39–40
Project Inclusion, 111–114
research approach to a general chemistry laboratory, 98–99
"Good" science and "bad" science, 107
Gordon Research Conference (GRC), 147, 149
The Grant Game: A Few Simple Rules/Several Bad Ideas, 157
Graphing calculator use, 125–126
Guest lecturers, 103

**H**

Hands-On Multimedia, 157
Hispanics, underrepresentation in science fields, 131
*How Scholars Trumped Teachers,* 149

**I**

ICONS program at the College of Santa Fe, 120–124
Inclusion
Project Inclusion, 111–114
of special education students in general academic classrooms, 170
Individuals with Disabilities Education Act (IDEA), 170
Informal assessments, 69
Innovations in College Chemistry Teaching, 149
Inquiry-Based Instruction workshop, 157
Inquiry-based learning
basis for, xvi–xvii
changing student attitudes about science and, 3–4
class size and, 5
facilitator role, 5–6
importance of, 2
inquiry definition, xvi
instructional strategies, xvii
labor-intensiveness of, 3
learning styles and, 6–7
logistics of inquiry, 3
program planning, 2–4
program standards and, 120
small group assignments and, 5
Institute for Conservation Studies (ICONS) at the College of Santa Fe, 120–124
Internet. *See* Web sites

Interrelationships among science education, government, national organizations and societies, and the private sector, 140

Interviews, observations, and focus groups, 63

Intuitive learners, 171

Iowa Chautauqua model, 43–45

## J

Journals as assessment tools, 69, 70–72, 76

## K

K-12 teachers, professional development, 26–28

Kakiemon Red dye, 113

## L

Laboratory experiences
    active learning and, 172
    content standards and, 97–99
    resources for revamping laboratories, 127–128

Large classes
    assessments in, 9–11
    content standards and, 97–98
    system standards and, 156, 157

"The Large Enrollment Classroom: Creating a Participatory Learning Environment," 156

The Large Enrollment Classroom: Creating an Inclusive Learning Community, 157

Large-scale survey work, 63

LEAD Center at the University of Wisconsin–Madison, 39–40

Learning communities
    ideal college science classroom, 16
    introductory biology example, 16–18
    shared planning and, 16
    uniqueness of individuals and, 15–16

Learning disabilities, 170–172

Learning environment
    developing experimental design skills of students, 13–15
    open-ended projects and, 13–15
    oral reports and, 15
    time, space, and resources challenges, 12–13

Learning styles, 6–7, 171–172

Learning through Evaluation, Adaptation, and Dissemination (LEAD) Center at the University of Wisconsin–Madison, 39–40

*The Liberal Art of Science: Agenda for Action,* 36

Librarians, 141

LICIL project, 125–126

Life science content standards, 100

Literature-based examinations, 84–86

Long Island Consortium for Interconnected Learning (LICIL) project, 125–126

Love Canal controversy, 109

## M

Math laboratories, 141

Mathematics, coordination with science education, 124–126

Medical National Board Part I Examination, 82

Mentoring, role in professional development, 51, 52

Merck & Co., Inc., support of undergraduate science education, 128–131

Merck/AAAS Undergraduate Science Research Program, 129

Merck Engineering & Technology Fellowship Program, 129–130

Mini-lecture concept, 72

Minorities. *See also specific groups*
    opportunity to learn, 131–133

Minority Access to Research Careers (MARC) program, 132–133

Minority Biomedical Research Support (MBRS) program, 133

Minority International Research Training (MIRT) program, 133

Minute papers, 10, 69, 156–157

Most confusing concept question, 10

Mount St. Mary's College
    program reviews, 117–119
    research approach to the general chemistry laboratory, 98–99

Multiple assessments, 8–11, 22, 61, 69, 75

## N

*A Nation at Risk: The Imperative for Educational Reform,* xii–xiii

National Aeronautics and Space Administration (NASA), 20–22

National Association for Research on Science Teaching (NARST), 147, 149

National Association of Biology Teachers (NABT), 143

National Commission on Excellence in Education (NCEE), xii, xiii

The National Faculty (TNF) programs, 45–46

National Governors' conference, national education meeting, xiii

National Institute of General Medical Sciences (NIGMS), 132

National Institutes of Health
    National Institute of General Medical Sciences, 132

National Research Council (NRC), 163, xiv
*National Science Education Standards*
    American Chemical Society task force on,
        160
    basis for, ix
    guiding principles, xv
    professional development standards view, 39
    responsibility for change, xvii
    role for higher education, 169, xviii
    scientific literacy goal, xiv–xv
    vision of, 1, 27–28, 42, 146, 164
National Science Foundation, 31, 37, 40, 125,
    128, 137, 143, 149, 169, 171, xiii
    Course and Curriculum grant, 78
    Shaping Future Faculty in Mathematics and
        the Sciences, 50
National Science Teachers Association (NSTA),
    143, xiii–xiv
Native Americans
    traditional basketmaking, 113
    underrepresentation in science fields, 131
NC State Index of Learning Styles
    Questionnaire, 6
New Traditions Chemistry Systemic Reform
    project, 40
Nonscience majors
    assessments and, 82–84
    content standards and, 101–103, 108–110
    program standards and, 116, 121–124
    system standards and, 151–155, 162, 165–
        167

## O

Occupational Safety and Health Administration
    (OSHA), Emergency Planning and
    Community Right to Know Act of 1986, 128
One-Minute papers, 10, 69, 156–157
"Online University Teaching Centers Across the
    World," 34
*On the Origin of Species*, 102
Open-ended projects, 13–15
Oral reports, 15

## P

Pedagogical content knowledge, 43
Pedagogical methods, 32
Peer coaching, 49
Peer review, 62, 63–65, 71
Performance-based assessments, 62, 65–68
Periodic sampling of intermediate materials, 62
"Periodicals Related to College Teaching," 34
Personal and social perspectives of science,
    107–110

The Pew Charitable Trusts, 50
Pharmacy programs, 130–131
*The Pharmacy Student Companion: Your Road
    Map to Pharmacy Education and Careers,* 130
Pharmacy Student Research Conference-Eastern
    Region, 130
Pharmacy Student Research Conference-
    Western Region, 130
Physical science content standards, 100
Portfolios as assessment tools, 69, 72–74, 76
Poster assignments, 82–84
Preparing Future Faculty (PFF) programs, 50–
    53, 149, 160
Preservice teachers, professional development,
    26, 30, 31–32, 38–39, 44, 46, 50–53
Pretests and posttests, 76–78
*Principles of Good Practice for Assessing Student
    Learning,* 58–59
Private industry, 142–143
Problem-Solving in Chemistry workshop, 157
Process education, 126
Professional and Organizational Development
    (POD) Network in Higher Education Web
    site, 34
Professional development standards
    administration's responsibility, 38
    analysis of an innovative college chemistry
        course, 39–40
    barriers to, 33
    community of learners concept, 43
    constructivist view of learning and, 41–42,
        48
    criteria for, 28
    educational seminars, 31–32
    equilibration concept, 42, 48
    faculty professional development centers,
        32–33
    faculty responsibility, 34
    focus of, 25, 30–31
    importance of, 26
    Internet and World Wide Web sites, 34
    Iowa Chautauqua model, 43–45
    for K-12 teachers, 26–28
    learning how to teach science, 30–34
    learning science content, 27–29
    learning to learn, 35–40
    mentor role, 52, 53
    mode variety, 34
    The National Faculty (TNF) programs, 45–
        46
    pedagogical content knowledge and, 43
    pedagogical methods, 32
    pertinence of, 25–26

planning professional development
programs, 41–54
preparing future faculty, 50–54
preservice teachers and, 26, 30, 31–32, 38–
39, 44, 46, 50–53
professional development process, 47–50
research-based change in teaching, 36–38
a science program for prospective teachers,
31–32
*Standards* view of, 39
successful program criteria, 26
summer activities, 38
teaching strategies and, 30
vision of the *Standards* and, 27–28, 42
workshops, 29, 33
Wyoming TRIAD (WyTRIAD), 46–50
Professional societies, 143. *See also specific
societies*
Program development
adaptation of Standard F for higher
education, 18–19
aligning courses, 20–22
curriculum reform and, 19
logistical arrangements of course delivery
and, 19
updating of course material, 19
Program reviews, 117–119
Program standards
continual evolution of programs, 117
mathematics coordination with science
education, 124–126
nonscience majors and, 116, 121–124
opportunity to learn, 131–133
process education and, 126
program assessment, 117
program consistency, 116–119
program of study description, 115
project-based approach to undergraduate
education, 120–124
resources to support the science program,
127–131
sequencing of courses and, 120
support for teachers, 134–138
Project 2061, 11, xiii
Project-based approach to undergraduate
education, 120–124
Project Inclusion, 111–114
Project Kaleidoscope (PKAL), 6, 135–138
Prompts, 88

## R

Reaction papers, 106
Reading specialists, 141

Real world experiences, 128
Research approach to the general chemistry
laboratory, 98–99
Resources
campus resources, 141–142
electronic resources, 141
resources for change, 158–161
for revamping laboratories, 127–128
for support of the science program, 127–
131
time, space, and resources challenges in the
learning environment, 12–13
Risk-benefit analysis, 108–109

## S

Sarstoon Temash protected area, 123
*Scholarship Reconsidered,* 148
Science
importance of, xiii
nature of, xii–xiii
*Science and Human Values,* 102
Science and technology, 104–107
Science as inquiry, 97–99
"Science Education Policies for Sustainable
Reform," 159
*Science for All Americans,* 36, 169, xiii
Science Process Skills Test (SCIPROS), 79–80
Scientific literacy, xiv–xv
Scientific method description, xii
SciTeKS project, 160
"Scope, Sequence, and Coordination of
Secondary School Science," viii–ix
Self-assessments, 9, 83–84
Senior citizens, 20–22
Sensory learners, 171
Shaping Future Faculty in Mathematics and the
Sciences, 50
"Shaping the Future" conference, 143
Shaping the Preparation of Future Science and
Mathematics Faculty initiative at Duquesne
University, 51–53
Small Group Instructional Diagnosis (SGID), 63
Social perspectives, content standards and,
105–110
Society for College Science Teachers (SCST),
143, 163
"State of the World," 106
Stereotype threat, 81
*Strategies for Success* series, 156–158
Strength in Numbers: Uniting the Fronts of
Higher Education symposium, 147–150
Structured Study Groups (SSGs), 63–68

*Student Assessment-as-Learning at Alverno College,* 88
Student Attitude Inventory (SAI), 79
Students with disabilities
    assessment bias and, 82
    learning disabilities, 170–171
    number of, 169–170
    teaching difficulties, 170–171
Suffolk Community College, nontraditional biology course for nonscience majors, 164–166
Summative evaluation, 60–61
SUNY-Farmingdale, graphing calculator use, 125–126
Sustainable Uses of Biological Resources project, 122
System standards
    campus resources subsystem, 141–142
    common values, 139
    communication requirement, 151
    constructing a local model system, 144–146
    constructivist view of learning and, 165–167
    coordination of science education policies, 150–155
    description, 139
    electronic resources, 142
    equitable policies, 161–162
    examples of components of science education's major subsystems, 144
    government subsystem, 143–144
    individual responsibility and, 164–167
    interrelationships among science education, government, national organizations and societies, and the private sector, 140
    national organizations and societies subsystem, 143
    nonscience majors and, 151–155, 162, 165–167
    policy review for unintended effects, 163
    the private sector subsystem, 142–143
    resources for change, 158–161
    science education from a postsecondary campus perspective, 142
    Strength in Numbers: Uniting the Fronts of Higher Education symposium, 147–150
    subsystems, 139, 141–146
    sustained policies, 155–158
    typical diagram of a university education system, 145
    uniformity and, 150
    vision of the *Standards* and, 146, 164
    workshops, 156–158

**T**
Teaching standards
    building learning communities, 15–18
    designing and managing the learning environment, 12–15
    guiding and facilitating learning, 4–7
    linking assessing, learning, and teaching, 8–11
    participating in program development, 18–22
    planning an inquiry-based science program, 2–4
    vision of the *Standards,* 1
Technology and science, 104–107
Tenure considerations, 146–147
Think-Pair-Share strategy, 156–157
Thinking learners, 171
Third International Mathematics and Science Study (TIMSS), viii
"Tragedy of the Commons," 103

**U**
UNCF-Merck Science Initiative, 129
Undergraduate Science Research Program, 129
University of Alaska, Fairbanks, 152–155
University of California, conference on implementing the *Standards,* 147
University of Colorado School of Pharmacy, Pharmacy Student Research Conference-Western Region, 130
University of Wisconsin–Madison Learning through Evaluation, Adaptation, and Dissemination (LEAD) Center, 39–40

**V**
Vision of the *Standards*
    professional development standards and, 27–28, 42
    system standards and, 146, 163
    teaching standards and, 1

**W**
Web sites
    ACS Education Web site, 160
    Professional and Organizational Development (POD) Network in Higher Education Web site, 34
West Virginia University School of Pharmacy, Pharmacy Student Research Conference-Eastern Region, 130
W.M. Keck/PKAL consultancy program, 135–138

Women
   assessment bias and, 81–82
   scientific contributions of, 112
Workshops
   content standards and, 98–99, 113
   professional development standards and, 29,
      33
   system standards, 156–158
World Watch Institute, 106
Writing laboratories, 141
Wyoming TRIAD (WyTRIAD), 46–50